NUCLEAR
MADNESS

ALSO BY HELEN CALDICOTT

Missile Envy
If You Love This Planet

HELEN CALDICOTT, M.D.

W. W. NORTON & COMPANY
NEW YORK/LONDON

Nuclear

What You Can Do

MADNESS

What You Can Do

REVISED EDITION

The text of this book is composed in Garamond 156
with the display set in Industria Solid
Book design by Chris Welch

ISBN 0-393-03603-0
ISBN 0-393-31011-6 (pbk)
W. W. Norton & Company, Inc., 500 Fifth Avenue, New York, N.Y. 10110
W. W. Norton & Company Ltd., 10 Coptic Street, London WC1A 1PU
1 2 3 4 5 6 7 8 9 0

To my children and children-in-law, Philip, Fiona, Penelope, Eric, and William; and to my grandchildren, Mikhael and Rachel

Contents

Acknowledgments

I thank the following people for the hours of advice and help given to update this book: Mary Olson, Captain James Bush, Mary Cunnane, and others who prefer to remain anonymous.

Introduction

Since I first wrote *Nuclear Madness* in 1978, many things have happened. In March 1979, Three Mile Island (TMI) melted down and became a household word for the horrors of nuclear power. Then in March 1986, Chernobyl blew its top and distributed radioactivity throughout Europe and indeed the Northern Hemisphere. As many as half a million people could be condemned to die of cancer or leukemia, or to be born genetically deformed, as a result of this catastrophe.

Australia elected a Labour Government in 1982, led by Prime Minister Bob Hawke, which overthrew its own anti-uranium policy enacted in the 1970s as a result of enlightened moves by the Australian Council of Trade Unions. But despite the fact that Australia exports uranium to nine countries now, we have no commercial nuclear power plants, and at this stage only a small, aging, very dirty research reactor on the outskirts of Sydney.[1] We grow nonradioactive food, which is one of our main export assets, but because the government is pro-uranium, it chooses not to advertise our nonradioactive food to a world that is becoming irrevocably polluted with long-lived radioactive isotopes, some of which are manufactured from our uranium. However, the Australian government is about to sabotage our clean food policy by agreeing to cooperate with the proposed Indonesian nuclear power program. Indonesia plans to construct up to 12 nuclear reactors in locations prone to earthquakes and volcanoes in the next 10–15 years. A nuclear meltdown could well contaminate much of the Australian land mass, and we will then be exporting radioactive food, to a radioactive Northern Hemisphere. Plans are being made to supply Australian uranium to Indonesian reactors.[2]

I returned to live in Australia in 1987, after fourteen years in the United States. During the intervening years I visited North America on lecture tours at least twice a year and have kept abreast of policies and changes in the nuclear scene. Amazing revelations have surfaced both in the United States and the Confederation of Independent States (CIS) over the last four years, exposing the most horrific radioactive mess contaminating the land of most states of the United States of America and many of the republics of the former Soviet Union.

Together, the nuclear establishments of the cold war warriors have engaged in the biggest cover-up in the history of the world. The legacy of cold war nuclear bomb production translates into a hot war of contaminated food, air, and water, both for us and for all future generations of humans, animals, and plants.

And to make matters worse, despite the horror and sense of repulsion evoked by Three Mile Island and Chernobyl, the nuclear power industry of the United States has been diligently working to reassure an uneasy public that nuclear power is clean, cheap, and safe.

One of its arguments is that no greenhouse gases are produced at nuclear reactors. But the truth is that the main global warming gas, carbon dioxide, is emitted at each step of the nuclear fuel chain, from uranium mining, milling, enrichment, fuel fabrication, construction of the reactor, transportation and storage of radioactive waste, and decommissioning of old reactors. So nuclear power adds to greenhouse warming as well as to radioactive pollution.

Nuclear power is now by far the most expensive form of electricity production, if one calculates the cumulative cost of taxpayer subsidies to the industry at each step of the fuel chain, from uranium mining right through to the storage of radioactive waste.

It is also obviously extremely unsafe, as opposed to the fallacious claims made by the nuclear industry in their eye-catching advertisements. This book enumerates the medical dangers at each step of the nuclear fuel chain.

The number of jobs created by an almost defunct industry are minute compared to the number of potential jobs that could be created by constructing massive solar and wind electricity farms, backfitting every building in the United States to solar power, and ensuring that all new buildings are heated, cooled, and powered by the sun and wind.

What a vision for a grand new world, and what a wonderful model for the developing world, which could be so inspired by this example that it might be helped to bypass the fossil fuel and nuclear era, thus reducing greenhouse warming and additional nuclear pollution.

But as nuclear power becomes more and more distasteful to U.S. citizens, General Electric, Westinghouse, and other nuclear companies are bandying their wares to other countries such as Indonesia, the Philippines, Thailand, Malaysia, Taiwan, China, and former Eastern Bloc States. No country has a solution to the storage of radioactive waste. Most pretend that it does not exist. They turn a blind eye to the problem, while the thermally hot, used fuel rods pulsate with radiation in their fuel pools, promising an uncertain and dangerous future for our children and all future generations.

Recent revelations sparked by the U.S. Secretary of Energy, Hazel O'Leary, have exposed Nazi-like radiation experiments which were performed on more than 1,000 innocent American citizens throughout the Cold War years.

From 1946 to 1956, 49 retarded boys at the Fernald State School in Waltham, Massachusetts, were fed radioactive iron and calcium in their breakfast cereal by scientists from Harvard and MIT without their parents' informed consent. A dose equivalent of 50 X-rays was incurred. Many of these boys have since died without any followup. The experiments were sponsored by the Atomic Energy Commission and Quaker Oats.

750 poor, pregnant women were given radioactive iron at a free prenatal clinic at Vanderbilt University in Nashville, exposing their fetuses to radiation levels thirty times above normal. At least three of the children died of cancer by the age of eleven.

As recently as 1971 over 100 prisoners in Oregon and Washington state prisons were paid $5 a month so that scientists could irradiate their testicles with extremely high doses of radiation. They all became temporarily sterile. They were later given vasectomies to "avoid the possibility of contaminating the general population with radiation-induced mutants," according to Dr. Carl Heller who ran the experiments. Catholics were exempted from the vasectomy.

At the University of Cincinnati College of Medicine in the 60s and 70s, indigent cancer patients were exposed to very high levels of radiation so they developed acute radiation illness. Nine of the first forty irradiated died within 38 days.

The Oak Ridge National Laboratories in Tennessee exposed nearly 500 patients with leukemia and other cancers to exceptionally high levels of radiation from radioactive cesium and cobalt, including a six-year-old boy.

Argonne National Laboratories outside Chicago injected 18 patients with plutonium between 1945 and 1947. All have since died.

Seven newborn babies, six of whom were black, were injected with radioactive iodine to see if it was absorbed by their thyroid glands.

Between 1960 and 1971, the Atomic Energy Commission irradiated 87 destitute cancer patients with hundreds of rads (battlefield doses) to determine what level of radiation would cause a soldier "to not follow orders, be dysfunctional or throw up in his face mask in an airplane." Nine died within twenty days of irradiation. These patients had an average IQ of 86.

An army study in the 1950s gave radioactive iodine to Eskimos and Indians in Alaska and to GIs stationed there to study thyroid function in cold weather.

NASA signed an agreement with the AEC in 1964 to perform radiation experiments on people to determine how much radiation from the sun would affect astronauts.

The Department of Energy is still engaged in scores of radiation experiments for its own research.

The CIA is not cooperating in the review ordered by Secretary O'Leary but, according to Mr. Steven Aftergood of the Federation of American Scientists, the CIA tests were "more secret and more lethal" than other departmental experiments.[3]

More is yet to be revealed in 32 million pages of documents about to be released by the DOE.

As well, the United States conducted 204 secret underground nuclear tests over 45 years, and at least 34 of them vented radiation into the atmosphere.[4]

Senator John Glenn recently revealed that the United States deliberately dropped radioactive material from aircraft or released it on the ground a dozen times after 1945. Eight of the tests were part of secret radiation weapons research in Tennessee and Utah; four other times, radiation was released to the atmosphere so that pilots could chase the fallout cloud to assess its mobility. People living downwind were irradiated. This was described as a "systematic radioactive warfare program" on the American people by Mr. Arjun Makhijani, president of the Institute for Energy and Environmental Research.[5]

Meanwhile, military nuclear waste in the form of spent fuel rods rusts and leaks in its cluster of cooling pools at Savannah River, South Carolina; Hanford, Washington; and Idaho National Engineering Laboratories near Idaho Springs. Radiation is leaking so fast that purification systems can barely keep up and workers have been exposed to sharply higher radiation levels recently.[6] And explosions of reprocessed high-level radioactive waste at eleven reprocessing facilities in six states may be imminent. Such an event would permanently contaminate surrounding areas of land and populations.[7]

Russia wants to dump more radioactive waste into the sea because of a catastrophic lack of storage space.[8]

And to cap off the saga of nuclear nightmares, a woman with acute myelogenous leukemia is suing Southern California Edison over the lack of safety precautions at the San Onofre nuclear power plant. She has six months to live and was contaminated by fuel "fleas," microscopic particles of radioactive material that escaped from tiny holes in sealed areas during her two years of employment as an inspector from 1985 to 1986. This is the first worker safety suit brought by a U.S. nuclear employee. There will be thousands more.[9]

NUCLEAR MADNESS

Chapter 1

Our Own
Worst Enemy

I am a child of the atomic age. I was six years old when American atomic bombs were deployed against the Japanese, and I have grown up with the fear of imminent annihilation by nuclear holocaust. At this writing, nuclear power has metastasized around the globe, with a total of 422 nuclear power plants worldwide and forty-five under construction. There are forty-four reactors in Japan with ten in production; fifty-six in France, with five in production (many are aging and need expensive repairs);[1] and thirty-seven in the United Kingdom, with one to be completed. The United States has 108 operating reactors, 20 are closed and 2 are under active construction; its nuclear industry plans to build 175 more over the next thirty-six years, if it can obtain the necessary funding. The nuclear facilities stand to inherit the earth.[2]

As a physician, I contend that nuclear technology threatens life on our planet with extinction. If present trends continue, the air we breathe, the food we eat, and the water we drink will soon be contaminated with enough radioactive pollutants to pose a potential health hazard far greater than any plague humanity has ever experienced. Unknowingly exposed to these radioactive poisons, some of us may be developing cancer right now. Others may be passing damaged genes, the basic chemical units that transmit hereditary characteristics, to future generations. And more of us will inevitably be affected unless we bring about a dramatic reversal of the world's pronuclear policies.

I learned about the carcinogenic (cancer-causing) and mutagenic (gene-altering) effects of nuclear radiation during my first year of medical school, in 1956. At that time, the United States and the Soviet Union were conducting atmospheric bomb tests. My genetics teacher did not connect his lectures on radiation to the fall out produced by these explosions, but the relationship became clear to me when newspapers reported that radioactive strontium 90, a by-product of atomic testing, had been carried around the world by high-level winds, deposited on the earth as nuclear fall out, and was being found in high concentration in cow's milk and in the deciduous teeth and, presumably, the bones of children. At the same time, the long-term medical consequences of radiation were just beginning to appear, in the form of an increased rate of leukemia among Japanese atomic bomb survivors.

Most people with whom I've spoken know very little about the medical hazards posed by nuclear radiation, although they have, over time, been sensitized to the risk of nuclear war. Atomic bomb anxiety was prevalent in the 1950s, and most Americans were acutely aware of the devastation that nuclear war implies. Children were constantly diving under their desks and crouching in school corridors during simulated nuclear attacks, and a film called "Duck and Cover" advocating such behavior was distributed widely by the U.S. government. The bomb shelter business boomed and the nation prepared a program of civil defense. Americans were justifiably scared. People recognized that nuclear disarmament could mean the difference between a secure future for them and their children, and no future at all.[3]

But during the 1960s, the American public became preoccupied with other matters: political assassination, the Civil Rights movement, the Vietnam War. The threat of nuclear holocaust was submerged by these more immediate problems. Only the Pentagon sustained its interest in nuclear development: in the race to keep ahead of the Soviet Union, America's military strategists continued to stockpile more and more nuclear weapons, while developing and refining delivery systems, satellites, and submarines.

In the early 1980s, the antinuclear movement led by Physicians for Social Responsibility revitalized peoples' concern about the medical implications of nuclear war; and by 1985, 75 percent to 80 percent of Americans opposed the nuclear arms race. Mikhail Gorbachev then ascended to the Soviet presidency, and as a result of his policies, the Berlin Wall was dismantled and Germany reunited, the Baltic States were set free, and Glasnost and Perestroika ended the Communist State. The cold war was in effect over, and both the United States and the former Soviet Union began dismantling their nuclear missiles.

In the 1990s, however, it is of the utmost urgency that we refocus our attention on the problems posed by nuclear technology, for we have entered and are rapidly passing through a new phase of the atomic age. Despite the fact that reactor technology is beset with hazardous shortcomings that threaten the health and well-being of the nations that employ it, nuclear power plants are still spreading throughout the world. Moreover, by making peaceful nuclear technology available to any nation wealthy enough to buy a nuclear reactor, we are inviting other countries to join the international "nuclear club" militar-

ily, as well as economically. Officially, there are five nations
with nuclear weapons: France, China, Great Britain, the
United States, and the Confederation of Independent States.
But unofficially, Israel, India, South Africa, and Pakistan defi-
nitely possess nuclear weapons, while Argentina, North and
South Korea, Japan, Iraq, Iran, Taiwan, and Brazil, among
others, have shown possible interest and have the capability to
manufacture nuclear weapons.[4] Any country owning a nuclear
reactor is capable of joining the global nuclear weapons club.

In view of the threat that nuclear technology poses to the
ecosphere, we must acknowledge that we have reached an
historic turning point. Thousands of tons of radioactive materials released by uranium mining, milling, enrichment, and by
routine operation of the plant, as well as accidental releases and
nuclear explosions, are now dispersing through the environment. The nuclear fuel itself is one million times more radioactive when it comes out of the reactor than when it went in. In
addition, the entire reactor becomes contaminated, and the
whole thing will eventually become radioactive waste. Nonbi-
odegradable, and some potent virtually forever, these toxic
nuclear materials will continue to accumulate, and eventually
their effects on the biosphere and on human beings will be
grave. Many people already have, and many more will begin to
develop and die of, cancer; or their reproductive genes will
mutate, resulting in an increased incidence of congenitally
deformed and diseased offspring, not just in the next generation, but for the rest of time. An all-out nuclear war, a remote
but still a viable possibility, would kill millions of people and
accelerate these biological hazards among the survivors. The
earth would be poisoned and laid waste, rendered uninhabitable for eons.

"But what can I do, as an individual?" is a refrain I hear
whenever I draw people's attention to the problems threatening our survival. Economic pressures and the frustration of

dealing with a biased government and unresponsive bureaucracy leave many people feeling helpless, although we did prove in the 1980s that individuals can alter world events.

My experience in Australia from 1971 to 1976 and later, taught me that democracy can be made to work—that by exerting electoral pressure, an aroused citizenry can still move its government to the side of morality and common sense. In fact, the momentum for movement in this direction can only originate in the heart and mind of the individual citizen. Moreover, it takes only one person to initiate the process, and that person may be politically naive and inexperienced, just as I was when I first spoke out.

As a pediatrician, I have devoted most of my professional life to working with children born with cystic fibrosis, the most common inherited childhood disease. The abnormal gene that carries this fatal disorder is found in one in twenty Caucasians; the disease's incidence is one in 1,600 live births. Watching my patients die of respiratory failure, and seeing other children in the wards die of leukemia and cancer, has motivated me to speak publicly and to write this book. Knowing that the incidence of congenital diseases and malignancies has and will continue to increase in direct ratio to the radioactive contaminants polluting our planet, I cannot remain silent.

My personal commitment to human survival was sparked when I read Bertrand Russell's autobiography. In Russell I found the moving example of a man who faced up to the dangers of the atomic age and, despite all odds, dedicated himself to ridding the earth of nuclear weapons. By 1962, his "ban the bomb" movement had culminated in the International Test Ban Treaty signed by the United States, Great Britain, and the Soviet Union; and the world briefly waxed hopeful that the superpowers would begin to disarm. When, in 1971, I discovered that France had been conducting atmospheric tests over its small colony of Mururoa in the South

Pacific since 1966, thus contravening the treaty inspired by Russell's work, I became indignant. I knew that when an atomic bomb explodes near the earth's surface, the mushroom cloud that billows into the sky carries particles of radioactive dust. Blown from west to east by stratospheric winds, these particles descend to the earth in rainfall and work their way through the soil and water into the food chain, eventually posing a serious threat to human life.

At the time few Australians knew of the testing or were aware of its inherent dangers. I decided that it was my duty as a physician to protest France's disregard for the health of my fellow Australians, and I began by writing a letter to a local newspaper. That letter generated some supportive correspondence, and a TV news program asked me to comment on the medical hazards posed by fall out. France had tested another nuclear device, and planned to detonate another four more in the next few months. Each time the French tested a bomb I appeared on television again, explaining the dangers of radiation. As the public became better informed, a movement to stop the French tests coalesced around the medical facts.

The informal campaign that grew up around these televised talks gained greater momentum when I exposed a secret government report, passed on to me by a sympathizer employed by the state government of South Australia, which confirmed that in 1971 a high level of radiation had been found in South Australian drinking water. In June 1972, government inspectors detected 1,860 picocuries of radiation per liter of rainwater, compared with a normal background radiation level of 50 picocuries and the "safe" maximum of 1,000 arbitrarily established by the International Commission of Radiological Protection. Shocked by the realization that their water supplies were now sufficiently contaminated to pose a genuine threat to them and their children, the citizens of Australia took action. Thousands joined marches held in Adelaide, Melbourne, Brisbane, and Sydney; a spate of editorials and statements by eminent scientists demanded an end to the testing; newspapers devoted full pages to angry letters from readers; dockworkers refused to load French ships; postal workers refused to deliver French mail; consumers boycotted French products. A Gallop poll showed that within a year 75 percent of the Australian public had grown opposed to the French tests. And on June 15, 1972, in response to enormous public pressure, the Australian government instructed its representatives to the United Nation's Stockholm Conference on the Environment to vote in favor of a resolution calling for an end to all nuclear testing.

Although the Australian government was beginning to shift its position in response to public demand, the French remained unmoved, and it was decided by the committee coordinating the groups working together to stop the French tests that a delegation of concerned citizens should fly to Tahiti to lodge a formal protest with the governor of France's South Sea islands. The group was to include churchmen, union representatives, and Dr. Jim Cairns, deputy leader of the Australian Labour Party. I was invited to go along in June. Moments before we were to depart, however, we learned that French officials in Tahiti had announced that our group would be denied the right to disembark. We then decided to go to Paris with our protest. Twenty-four hours later, Jim, Ken Newcombe (leader of the Union of Australian Students), and I arrived in London, where we addressed a large crowd of Australians and New Zealanders in Hyde Park and delivered a letter of protest to the French embassy. Unfortunately, Britain's parlimentarians were far more concerned with securing England's entry into the Common Market than with hearing about the radioactive fall out then contaminating Australia, and Prime Minister Harold Wilson declined to meet with us. We were not surprised when we met with an equally cold political reception in Paris the next day. The bureaucrats with whom we exchanged opinions were closed to our arguments; they insisted that France needed its own *force du frappe* and that it would under no circumstances yield to public pressure. Stubbornly claiming that their nuclear tests were harmless, these professional civil servants nevertheless conceded that they would never consider conducting such tests in the Mediterranean: *"Mon Dieu,* there are too many people there!" We returned to Australia feeling that we had presented our case as forcefully as we could, but sorely disappointed by the arrogant disdain with which we were received.

In November, however, the momentum began to shift dramatically in our favor: a conference of Pacific nations convened by Australia and New Zealand spearheaded a UN vote to outlaw nuclear testing. The French countered with an announcement that 1973 would bring a new series of tests, including a 1-megaton hydrogen bomb blast over Mururoa Atoll. One month later, an enraged Australian public elected the Australian Labour Party, which opposed the action of the French, into office. Calling for an injunction to compel the French to desist from further atmospheric testing in the Pacific, the newly elected Australian government, and the government of New Zealand, took France before the International Court of Justice in the Hague. Although the court's decision was disappointingly equivocal, France finally backed down in the face of world opinion and announced that it would restrict its testing to underground sites.

In our struggle to put an end to atmospheric nuclear testing, we had demonstrated that one voice was all it took to raise a warning call, and that once enough other voices joined in, that call would be heard around the world.

My experience in the struggle I have just described, and in others in Australia and in the United States, has taught me many things: first and perhaps most important, that we can no longer afford to entrust our lives, and the lives and health of future generations, to politicians, bureaucrats, "experts," or scientific specialists, because all too often their objectivity is compromised. Most government officials are shockingly uninformed about the medical implications of nuclear power and atomic warfare, and yet they daily make life and death decisions in regard to these issues. Some responsibility for this ignorance must rest with my medical colleagues: too many of us are reluctant to look beyond our research laboratories or hospital corridors, and too many of us have remained silent about the medical hazards of nuclear technology and the radiation it produces, despite the fact that we acknowledge such radiation to be a certain cause of cancer and genetic disease.

My experience has also taught me that the survival of our species depends upon each individual. The controversy surrounding nuclear fission is the most important issue that all societies and the world at large have ever faced. A national and international debate on this subject is long overdue, and the participation of each individual will determine its outcome. We must begin by first of all learning as much as we can about the critical health hazards involved, because what we don't know about these dangers may kill us:

- The world's major military powers have built tens of thousands of atomic bombs powerful enough to kill the world's inhabitants several times over, and, despite the fact that nuclear disarmament is occurring between the superpowers, it is a very slow process.
- Each 1000-megawatt nuclear reactor contains as much long-lived radioactive material ("fall out") as would be produced by one thousand Hiroshima-sized bombs. A "meltdown" (in which fissioning nuclear fuel overheats and melts, penetrating the steel and concrete structures that encase it) could release a reactor's radioactive con-

tents into the atmosphere killing hundreds of thousands of people, depending upon the wind direction and population density, and contaminating thousands of square miles.

• Each operating reactor daily releases carcinogenic and mutagenic effluent. These radioactive materials raise the level of background radiation to which we are constantly exposed, increasing our risk of developing cancer and genetic disease.

• Each reactor annually produces tons of radioactive waste, some of which remains dangerous for more than 500,000 years. No permanant fail-safe method of containment or storage has yet been found for them, despite millions of dollars spent during four decades of research. Despite recent proposals, there is good reason to suspect that we may never develop safe methods of containment and long-term storage. Since there is currently no other place to put the used irradiated fuel/high-level waste, it is today sitting at each reactor site; thus, the cumulative total of the bulk of the radiation generated is accumulating on sites that are usually located on fresh water supplies.

• Each reactor annually produces approximately five hundred pounds of plutonium. Dangerous for at least 500,000 years, this toxic substance poses a threat to public health that cannot be overemphasized. Present in nature in only minute amounts, plutonium is one of the deadliest substances known. In addition, it is the basic raw material needed for the fabrication of atomic bombs, and each reactor yearly produces enough to make forty such weapons. Thus, "peaceful" nuclear power production is synonymous with nuclear weapons proliferation. American, French, German, British, Swedish, Japanese, and Canadian sales of reactor technology abroad guarantees that by the end of the century dozens of countries will possess enough nuclear material to manufacture bombs of their own. Moreover, the "plutonium economy," which has been adopted by Japan and still pursued by some in the U.S. nuclear industry and its supporters in government, presents the disturbing probability that terrorist groups will construct atomic bombs from stolen nuclear materials, or that criminals will divert such material for radioactive blackmail. Because of this terrorist threat we may find ourselves living in a police state designed to minimize unauthorized access to such nuclear materials.

Most early developers of nuclear energy explored its potential fifty years ago to produce bombs that would inflict unprecedented damage. Seven years after the United States tested two such weapons on the populations of Hiroshima and Nagasaki in 1945, the collective guilt generated by the deaths of some 350,000 Japanese civilians prompted the American government to advocate a new policy: the "peaceful use of atomic energy" to produce "safe, clean electricity," a form of power touted as being "too cheap to meter." Together, industry and government leaders decided that nuclear power would become the energy source of the future. Today, thirty-eight years later, that prospect still threatens the well-being of many nations and of the world.

Despite Three Mile Island and Chernobyl, the global nuclear "priesthood" is still intent on building new reactors. One hundred and seventy five are planned for the United States alone over the next several decades.

One need not be a scientist or nuclear engineer to take part in this important debate; in fact, an over-specialized approach tends to confuse the issue. The basic questions involved ultimately go beyond the technical problems related to reactor safety and radioactive waste management. It is now obvious that we cannot have faith in the infallibility of human beings to administer this unforgiving technology. Even if it is possible to isolate nuclear waste from our environment, how confident can we be in our ability to control the actions of fanatics or criminals? How can we assure the longevity of the social institutions responsible for perpetuating that isolation? And what moral right do we have to burden our progeny with this poisonous legacy, let alone to produce more? Finally, we must confront the philosophical issue at the heart of the crisis: Do we, as a species, possess the wisdom that the intelligent use of nuclear power demands? This question is, of course, redundant because nuclear power is the one industry where we *can't afford* to be fallible.

From a purely medical point of view, there really is no controversy: the commerical and military technologies we have developed to release the energy of the nucleus impose unacceptable risks to health and life. As a physician, I consider it my responsibility to preserve and further life. Thus, as a doctor, as well as a mother and a world citizen, I wish to practice the ultimate form of preventive medicine by ridding the earth of these technologies that propagate disease, suffering, and death.

Chapter 2

Radiation

Radiation, the particles and waves emitted by unstable elements, has saved the lives of thousands of people when used to diagnose and treat disease. But little more than thirty years after its discovery in the late 1890s scientists began to find that radiation had a dual nature: it could kill as well as cure. Working with primitive, high-dose X-ray machines, many early roentgenologists died of radiation burns and cancer. Well-known for their pioneering work with radium, Marie Curie and her daughter Irene died of leukemia. Studies conducted over the past forty years have shown that many people irradiated in infancy and childhood for such minor maladies as acne, enlarged thymus, bronchitis, ringworm, tonsillitis, and adenoids have developed cancers of the thyroid, salivary glands, brain, pharynx, and larynx up to thirty years later. Studies of uranium miners and people engaged in commercial activities, as well as Japanese survivors of atomic explosions, have yielded enough evidence to demonstrate beyond doubt that cancer of the blood, lung, thyroid, breast, stomach, lymph glands, and bone occur in human beings as a result of exposure to radiation. Today, therefore, it is an accepted medical fact that radiation causes cancer.

To understand the dangers posed by nuclear power generation, nuclear weapons production, and nuclear warfare, we must acquire a basic knowledge of the nature of radiation and its biological impact on human body cells.

All matter is composed of elements, and the smallest particle of an element is an atom. Each atom has a central nucleus consisting for the most part of protons (particles with mass and positive electric charge). The number of protons in the nucleus gives us the element's "atomic number": the sum total of both the protons and neutrons gives us the element's "atomic weight." All the atoms of a given element have the same atomic number, but because some atoms contain more neutrons than others, not all of an element's atoms have the same atomic weight. Atoms of the same element with different atomic weights are called "isotopes." Uranium, for example, with an atomic number of 92, appears in nature in two forms: uranium 235 and uranium 238.

All elements with an atomic number of 83 or more are unstable or "radioactive," which means that their atoms can spontaneously eject—or "radiate"—particles and energy waves from their nuclei. This emission process, during which an element disintegrates into other nuclear forms, is referred to as "radioactive decay," and the rate at which it proceeds is calculated in terms of "half-life." The half-life of an element is the period of time it takes for the radioactivity of any amount of that element to be reduced by half. The half-life of strontium 90, for example, is twenty-eight years. Starting with one pound of strontium 90, in twenty-eight years there will be half a pound of radioactive material, in twenty-eight more years there will be a quarter of a pound, in twenty-eight more years there will be one-eighth of a pound. After approximately 560 years or twenty half-lives, the radioactivity of a given sample of strontium 90 will be reduced to one millionth of its original potency.

In the course of this decay, atoms give off three major forms of radiation: alpha, beta, and gamma, named after the first three letters of the Greek alphabet.

The equivalent of a helium nucleus, an alpha particle consists of two protons and two neutrons. Because of its relatively great size and weight, such a particle can be stopped by a sheet of paper, tends to lose momentum quickly, and can penetrate only short distances into matter; nevertheless, it is very energetic, and if it is moving fast enough when it comes in contact with a living body cell, it can burst through the cell wall and do serious damage to the interior. In fact, for the same amount of total energy delivered, alpha radiation has greater biological effects than any other form of radiation. Recent studies show the alpha impact on chromosomes to be one thousand times greater than gamma radiation.[1]

Almost two thousand times smaller than an alpha particle, a beta particle, when negatively charged, is identical to an electron. Emitted by the nucleus, beta particles can penetrate matter much further than alphas: they can travel through a number of body cells before they lose energy and come to a stop.

Gamma radiation, electromagnetic energy waves emitted by the nucleus of a radioactive substance, has the greatest penetrating power and often accompanies alpha and beta emission. X rays are similar to gamma rays.

Radiation is insidious, because it cannot be detected by the senses. We are not biologically equipped to feel its power, or see, hear, touch, or smell it. Yet gamma radiation can penetrate our bodies if we are exposed to radioactive substances. Beta particles can pass through the skin to damage living cells, although, like alpha particles, which are unable to penetrate this barrier, their most serious and irreparable damage is done when we ingest food or water—or inhale air—contaminated with particles of radioactive matter.

Radiation harms us by ionizing—that is, altering the electric charge of—the atoms and molecules composing our body cells. Whether the effects of this ionizing are manifest within hours or over a period of years usually depends on the amount of exposure, measured in terms of rem (roentgen equivalent man) units. Nevertheless, even the smallest dose (measured in millirems) can affect us, for the effects of radiation are cumulative. If we receive several small amounts of radiation over time, the long-term biological effect (cancer, leukemia, genetic injury) is almost certainly similar to receiving a large dose all at once.

A very high dose of ionizing radiation (say, of three thousand rems or more) causes acute encephalopathic syndrome— an effect scientists sought when they designed a "neutron bomb" to be used against invading forces. The explosion of such a bomb will leave buildings intact (although they may remain radioactive for years); what is destroyed is the human brain and nervous tissue. Within forty-eight hours of exposure the brain cells will swell and enlarge, producing increased pressure inside the skull. Confusion, delerium, stupor, psychosis, ataxia (the loss of neurological control of the muscles), and fever result; there follows a period of lucidity, then sudden death.

A dose of six hundred rems or more produces acute radiation sickness. Thousands of Japanese A-bomb victims died from this sickness within two weeks of the bomb explosions in 1945. Such exposure to radiation kills all actively dividing cells in the body: hair falls out, skin is sloughed off in big ulcers, vomiting and diarrhea occur; and then, as the white blood cells and platelets die, victims expire of infection and/or massive hemorrhage.

Lower doses of radiation can cause abnormalities of the immune system and can also cause leukemia five to ten years after exposure; cancer, twelve to sixty years later; and genetic diseases and congenital anomalies in future generations.

Of all the creatures on earth, human beings have been found to be one of the most susceptible to the carcinogenic effects of radiation (because their cells are rapidly dividing, fetuses, infants, and young children are the most sensitive to radiation's effects). One of the modern era's most dreaded killer diseases, cancer is like a parasitic organism, often causing slow and painful death. It is estimated by medical authorities that one in three people in the western world will contract the disease at some point.

The mechanism by which radiation causes cancer is not completely understood. It is currently believed, however, that it involves damage to the genes. Our bodies are made up of billions of cells. Inside each cell is a nucleus, and inside the nucleus are long bead-like strings known as chromosomes. These strings are DNA molecules, sequences of which are specific genes. Genes control every aspect of the individual's hereditary characteristics: hair color, eye color, personality factors, brain development, and so forth. Half of one's genes are inherited from one's mother, half from one's father.

Genes also control cellular activities, and within every cell there is thought to be a regulatory gene that controls the cell's rate of division. If our bodies are gamma-irradiated from the exterior, or if we inhale a particle of radioactive matter into our lungs and one of its atoms emits an alpha or beta particle, this radiation can collide with a regulatory gene and chemically damage it, sometimes killing the cell, sometimes leaving it alive. The surviving cell continues to function normally, until one day, five to sixty years later (i.e., after the "latent period" of carcinogenesis), instead of dividing to produce two new cells, it goes berserk and manufactures billions of identically

damaged cells. This type of growth, which leads to the formation of a tumor, is called cancer.

Cancer cells often break from the main mass of tumor,
enter the blood or lymph vessels, and travel to other organs.
Here again, they will divide uncontrollably to form new tumors. Because they are more aggressive than normal body
cells, cancer cells utilize the body's nutrients, causing normal
tissues to waste away and die.

In addition to giving rise to cancer, radiation also causes
genetic mutations—sudden changes in the inheritable characteristics of an organism. In 1927, Dr. H. J. Muller was awarded
the Nobel Prize for his discovery that X-irradiation causes an
increase in the number of such mutations in fruit flies. Muller's
findings have since been confirmed by many other researchers.
The genes and chromosomes of the scores of animals and
plants tested have been found to be vulnerable to radiation.
They, too, develop cancer and genetic diseases when exposed
to radioactive materials. The reproductive organs of human
beings are believed to be equally susceptible. Moreover, the
number of mutations had been shown to be in direct ratio to
the total amount of radiation exposure to the gonads, whether
that exposure be a single large dose or many very small ones.

A mutation occurs whenever a gene is chemically or struc-
turally changed. Some body cells die or become cancerous
when they are mutated; others survive without noticeable
changes. A genetically mutated sperm or egg cell may survive
free of cancer but can seriously damage the offspring to which
it gives rise.

There are two kinds of genes: dominant and recessive. To
clarify the difference, let us consider the example of eye color.
Each characteristic is determined by a pair of genes (one
member of the pair coming from the mother, one from the
father). The gene for brown eyes is a strong or dominant gene.
The gene for blue eyes is weak or recessive. A child who

inherits two brown-eyed genes will be born with brown eyes. The child with one brown-eyed gene and one blue-eyed gene will still have brown eyes, because the brown-eyed gene is dominant. The only way to get blue eyes is to inherit two blue-eyed genes.

A child formed from an egg or sperm cell mutated by radiation in a dominant way will show the results of that mutation. It may spontaneously abort or, if it survives pregnancy, it may turn out to be a sickly, deformed individual with a shortened lifespan. If this person then reproduces, statistically half his or her children will inherit the dominant gene and its deformities. Approximately five hundred such dominant genetic diseases have been identified. A typical example is achondroplastic dwarfism: individuals suffering from this disease are born with abnormal bones, resulting in short arms and legs and a relatively large head.

A radiation-induced recessive mutation might not make itself immediately apparent. A child might seem normal but carry the deleterious gene and pass it on to the next generation. Since the disease caused by a recessive gene will not manifest unless a child inherits the gene from both parents, it might not show up for generations. Diabetes, muscular dystrophy, hemophilia, certain forms of mental retardation, and cystic fibrosis are among the one thousand five hundred recessive genetic diseases now known.

Radiation can also cause chromosomal breakage in a sperm or egg cell, leading to seriously deformed offspring. One disease associated with chromosomal damage is mongolism, or Down's syndrome.

Deformities can also occur even when the sperm and egg cell are genetically normal, if radiation kills specific cells in the developing embryo during the first three months of intrauterine life. If a cell destined to form the septum of the heart is killed, a baby may be born with a hole in its heart. Such intrauterine damage, known as teratogenesis, can produce deformities similar to those caused by the drug thalidomide. Also, there is an increased incidence of childhood leukemia if the fetus is exposed to radiation in utero.

We are all exposed to background radiation in the form of the natural, "background" radiation to which the earth has been subject for billions of years. When the ozone layer of the atmosphere was thinner, ultraviolet rays from the sun and cosmic rays from outer space—two natural forms of radiation—streamed in unhampered to cause genetic mutations in every species. As a result of this and a complex of other forces, the simple single-celled organisms found in the ocean evolved gradually into more complex creatures adapted to living in the sea, on land, and in the air.

As the ozone layer accumulated and became more dense, these multicellular organisms were protected from damaging solar radiation, and eventually the human species developed, with its highly specialized brain. Strong or beneficial mutations prevailed, while the detrimental mutations died out. Almost all geneticists currently believe that humanity has reached an evolutionary peak in the number of beneficial mutations that the species can undergo; most genetic mutations are therefore thought to be detrimental, causing disease and deformity.

But background radiation continues to affect us. The protection of the ozone layer is now threatened, because chlorofluorocarbon (CFC) gases used in refrigerators, air conditioners, furniture manufacture, industry, and spray cans are damaging ozone in the upper atmosphere, allowing more ultraviolet and cosmic rays to reach the earth's surface. Consequently, the incidence of skin cancer and malignant melanoma is increasing dramatically, particularly in the Southern but also the Northern Hemispheres. Ultraviolet light also causes cataracts and blindness in humans and animals, and plants are themselves damaged by high levels of cosmic radiation.

Background radiation also comes from other natural sources, such as radium and radon, potassium 40, and carbon 14 present in rocks, air, and our own body cells. In Kerala, India, an abnormally high level of radioactive thorium found in the local soil is believed to be responsible for a high incidence of mongolism, mental retardation, and other congenital abnormalities. The average level of background radiation exposure for people living in both the Northern and Southern Hemispheres amounts to approximately one hundred millirems per year (six rems in sixty years). Although the exact percentage is unknown this radiation is thought to be responsible for a portion of all the cancers and genetic disorders afflicting us today.

Equally hazardous to our health is the human-made radiation to which most of us are exposed. Human-made radiation, too, can initiate cancer and genetic mutation. How does it reach us?

Medical X rays are the most prevalent source of radiation for the general public today. Since the effect of each dose is cumulative, each exposure carries with it an additional carcinogenic and mutagenic risk. Thus it is imperative that doctors, dentists, chiropractors, and patients alike take into account the potentially harmful effects of those rays, so that only those X rays deemed absolutely necessary are performed. In 1965, Dr. Karl Morgan, a noted health physicist, formally associated with the Atomic Energy Commission, estimated that 40–50 percent of all medical X rays were unnecessary and that patients were receiving twice the necessary exposure for X-ray procedures at that time. Dr. Morgan fought the unnecessary use of mass chest X-ray programs for twenty years, and this campaign dramatically reduced those procedures. Now he claims that compared to 400 percent unnecessary exposure in those days, we are down to 230 percent more exposure than is needed because of high insurance requirements and litigative medicine. Lawyers should remove themselves from the practice of medicine.[2]

Nuclear power production and the processes employed in the manufacture of nuclear weapons are responsible for generating billions upon billions of NEW radioactive atoms and molecules, and these are the second most prevalent sources of public exposure today. The difference is that you can turn X rays off, but radioactive waste lasts forever—the vast bulk of the POTENTIAL exposure for humans emanates from nuclear fission.

These fission processes result in the manufacture of hundreds of radioactive elements, which are already starting to contaminate the food chain. The radioactive material finds its way into rivers, lakes, and oceans, where it is eaten by fish, incorporated into their biochemical systems, and concentrated in their bodies thousands of times. Contaminated water is taken up by grass and other vegetation; the radioactive elements are concentrated again. Cows grazing on contaminated grass further concentrate the radiation and eventually pass the contamination onto us, in the form of milk or meat. Vast quantities of this man-made radiation lies buried in soils, migrating into aquifers, rivers, lakes, and oceans. It is routinely discharged into the air we breathe, and it sits in storage pools and huge storage tanks, eventually to contaminate the food chains for the rest of time.

Prominent among the radioactive elements manufactured in the production of nuclear power and atomic weapons are the beta and gamma emitters iodine 131, strontium 90, and cesium 137. Iodine has a half-life of eight days. Both this element and strontium 90 travel up the food chain, and, when ingested by humans, are absorbed through the bowel wall. Iodine 131 migrates in the blood to the thyroid gland and may cause cancer there twelve to fifty years later; strontium 90, with a half-life of twenty-eight years, chemically resembles calcium and is incorporated into bone tissue, where it may lead to leukemia and osteogenic sarcoma (a malignant bone tumor). Cesium 137, with a half-life of thirty years, concentrates in animal muscle and fish; ingested by humans, it deposits in body muscles and irradiates the muscle cells and nearby organs. Significantly, X rays can only damage the patient as they pass through the body, but radioactive waste remains in the body for many years consistently irradiating surrounding cells.

Whether natural or man-made, all radiation is dangerous. There is no "safe" amount of radioactive material or dose of radiation. Why? Because by virtue of the nature of the biological damage done by radiation, it takes only one radioactive atom, one cell, and one gene to initiate the cancer or mutation cycle. Any exposure at all, therefore, constitutes a serious gamble with the mechanisms of life.

Today almost all geneticists agree that there is no dose of radiation so low that it produces no mutations at all. Thus, even small amounts of background radiation are believed to have the potential for genetic effects.

Similarly, there is no disagreement among scientists that large doses of ionizing radiation cause a variety of different forms of cancer. Starting fifteen years after the explosions, the incidence of cancers of the stomach, ovary, breast, bowel, lung, bone, and thyroid doubled among Japan's bomb survivors. Approximately five years after the nuclear attack on Hiroshima, an epidemic of leukemia occurred that within ten years reached a level of incidence thirty times higher than among the nonexposed population.

The direct relation between cancer and even minute amounts of radiation has best been demonstrated by the British epidemiologist Dr. Alice Stewart, who found that only one diagnostic X ray to the pregnant abdomen increases the risk of leukemia in the offspring by 40 percent.

Every medical textbook dealing with the effects of radiation warns that there is no safe level of exposure. Nevertheless, the nuclear industry and government regulatory agencies have established what they claim to be "safe" doses for workers and the general public, drawing support from scientists who believe that there is a threshold below which low doses of ionizing radiation may in fact be harmless. This claim is dangerously misleading and, I believe, incorrect. The International Commission on Radiological Protection (ICRP) originally proposed "allowable" levels of exposure for use by the industry, but not without conceding that these may not be truly safe. Rather, it accorded priority to the expedient promotion of nuclear power. As the ICRP noted in its 1966 recommendations (Document #2), "This limitation necessarily involves a compromise between deleterious effects and social benefits . . . It is felt that this level provides reasonable latitude for the expansion of atomic energy programs into the foreseeable future. It should be emphasized that the limit may not in fact represent the proper balance between possible harm and probable benefit."

New data analyzed by physicists examining the Hiroshima and Nagasaki explosions suggest that radiation is far more deleterious than original calculations predicted. In fact, the fifth report of the Committee of the Biological Effects of Ionizing Radiation of the National Academy of Sciences, *BEIR FIVE,* now suggests that allowable radiation exposure limits be lowered three to five times. Others, such as eminent radiation expert Dr. John Gofman, suggest that they be lowered sixteen to thirty times. These new calculations will have enormous ramifications upon the medical, radiological, and nuclear power industries. They have yet to be adopted by the relevant government departments and implemented.[3]

The truth is that we are still courting catastrophe. The permissive radiation policy still supported by the American and other governments, in effect, turns us into guinea pigs in an experiment to determine how much radioactive material can be released into the environment before major epidemics of cancer, leukemia, and genetic abnormalities take their toll. The "experts" stand ready to count victims BEFORE they take remedial action. Meanwhile, the burden remains on the public to prove that the nuclear industry is hazardous rather than on the industry to prove that it is truly safe.

Today's safety standards have already been shown by several studies to be dangerously high. When investigators of low-dose ionizing radiation revealed that levels of radiation lower than those permitted were causing cancer, government agencies attempted to suppress their findings.

For instance, in February 1978, a classic case was vindicated at a hearing before the U.S. House Subcommittee on Health and the Environment. In 1964, the United States Energy Research and Development Administration (ERDA) had funded a study to be conducted by Dr. Thomas Mancuso, a physician and professor in the Public Health Department at the University of Pittsburgh. Its purpose was to determine whether low-level radiation induced any discernible biological effects in the nuclear workers at two of the oldest and largest atomic reactors in the United States, the facilities at Hanford, Washington, and Oak Ridge, Tennessee.

Dr. Mancuso's study was one of the broadest industrial epidemiological studies ever undertaken. Over a ten-year period, he studied one million files and compiled data from the death certificates of 3,710 former atomic power workers. Because of the long latency period of carcinogenisis, his first results were negative; that is, he did not find an incidence of cancer higher than the norms for the general public. In 1974, the Atomic Energy Commission (AEC) began pressuring him to publish his findings. The AEC wanted to use Mancuso's report to refute an independent study conducted by Dr. Samuel Milham of the Washington State Health Department; after reviewing 300,000 case histories at Hanford, Dr. Milham had found that there was indeed a high rate of cancer among former employees there.

Dr. Mancuso refused, claiming that his statistics were incomplete and that he needed more time. In 1975, the AEC informed him that his funding would be terminated as of July 1977 and demanded that he surrender his data to the Oak Ridge laboratories at that time. Mancuso took advantage of the interim to call in epidemiologist Stewart and her associate, Dr. George Kneale, a biostatistician, and together the three scrupulously studied the material on the Hanford workers. They came up with results similar to Milham's: a 6 to 7 percent increase in radiation-related cancer deaths among Hanford workers, indicating that the disease is distinctly related to radiation exposure at today's "acceptable" levels.

In fact, Dr. Mancuso discovered that the radiation "doubling dose" (that is, the dose at which the incidence of a disease is doubled) is 3.6 rads* per lifetime for bone marrow cancer, and stands at 33–38 rads per lifetime for other forms of cancer. Previous estimates, based on atomic bomb survivor data and human X-ray research, had set these doses at 100 rads for leukemia and 300–400 rads for solid cancer induction.

Today every civilian nuclear worker is allowed a radiation dose of 5 rads per year: that is, workers may be exposed to doubling doses for leukemia each year, and for cancer every seven to seven and a half years.

Dr. Thomas Najarian's study of Portsmouth, New Hampshire, nuclear submarine workers confirmed Dr. Mancuso's estimates. Such figures suggest that the nuclear industry will

---

*The rad (radiation absorbed dose) is a measure of radiation absorbed by a target expressed as the amount of energy (in ergs) per gram of absorbing material. The rem is a measure of the number of rads absorbed by a target multiplied by the relative biological destructiveness of the given type of radiation. In this case they are almost identical.

have to start hiring workers who are over sixty years of age—
so that they will not live long enough to develop malignancies!

The increasing exposure to radiation of workers and the
general public by the nuclear industries implies tragedy for
many human beings. Increasing numbers of people will have
to deal with cancer, or, perhaps more painful still, deformed or
diseased offspring.

In 1969, Dr. John Gofman and Dr. Arthur Tamplin, scientists formerly with the AEC's Lawrence Livermore Radiation
Laboratory, announced that if all Americans were annually
exposed to the official allowable doses of 170 millirems of
radiation (the equivalent of six chest X rays a year) over and
above background levels, there would be an increase of 32,000
to 300,000 deaths from cancer each year. As we will see, these
estimates are too low because of new data on the biological
effects of radiation exposure and because the "allowable" dose
from man-made radiation has now been increased to 100
millirems in the air, 100, via water, and up to 500 millirems "if
necessary," i.e., if there is an accidental release of radiation
from a nuclear reactor. Also the Nuclear Regulatory Commission (NRC) in their wisdom have averaged out high organ
doses (e.g. of iodine 131 to the thyroid) over the whole body.
This is an erroneous calculation because the thyroid in reality
still receives its original high dose. This is called "effective dose
equivalent," meaning in effect that people can receive even
higher doses than before and be at greater risk for cancer
induction. (Personal communication with Mary Olson from
the Nuclear Information and Resource Service, Washington,
D.C.)

It is difficult to predict how many mutated children will be
born in the world as a result of nuclear power and weapons
production, or what the nature of their defects will be. But it
is indisputable that the mutation rate will rise—perhaps far
higher than we would care to contemplate. The massive quan-

he was negotiating the sale of Australian uranium with the Shah of Iran.

Dr. Cairns was a politician in whom many people placed their trust. I could not understand why he would disregard the medical dangers associated with the mining of uranium and the health hazards inherent in the nuclear fuel cycle. In discussing the matter with him, however, I realized he was not even aware of the fact that the uranium-fission process produces plutonium, the material used to fuel atomic bombs, or that nuclear waste posed devastating health hazards.

This ignorance, unfortunately, is not unique. Over the past two decades, I have met many other politicians—in Australia, France, England, Ireland, Cuba, Czechoslovakia, Canada, the former Soviet Union, and the United States, who were equally uninformed. Nuclear power had been with us for more than three decades. Our public servants should have taken the initiative long ago to educate themselves on this life and death issue. Is it not their moral responsibility to do so?

The front end of the fuel cycle—the mining, milling, and enrichment of uranium—is dangerous, because at every step uranium decays into radioactive by-products that pose a threat both to workers in the nuclear industry and to the public at large.

MINING

In the United States, most uranium ore is mined in the Colorado Plateau and the Wyoming Basin. The mining is done by private companies, but through the Department of Energy (DOE) the taxpayers' money pays for the ore. During the process, two highly carcinogenic radioactive substances are released: radium and radon. Radium, an alpha emitter with a half-life of 1,600 years, is a decay product of uranium that is found in uranium ore. If particles of dust from uranium mines are swallowed, the radium is absorbed by the intestine and carried to the bone, where it can cause leukemia or bone cancer. Years ago, workers in watchmaking factories, predominantly women, painted the numbers on luminescent watch and clock faces using a radium-based paint. To make the figures more precise, many of them licked the tips of the paint brushes, swallowing relatively large amounts of radium.

A startlingly high percentage later died of bone cancer, leukemia, and acute radiation effects.

Radon, a gas, is a radioactive daughter isotope of thorium, which is a daughter product of uranium. Inhaled, it can cause lung cancer. Changes in mining safety standards to protect against radon have now been implemented, but 20 to 50 percent of the American, German, and Canadian miners working under conditions in the past have already died, or will die, from cancer.

Of the 1,500 Navajo men recruited in the 1940s through the 1960s from a simple farming life to mine uranium at Cove and Red Valley, Arizona, 1,112 miners or their families have filed for government compensation related to lung cancer and other radiation-induced diseases. Of these, 328 claims have been approved, 121 denied, and 663 are pending. Each is eligible for only $100,000 if successful. In 1949, the Atomic Energy Commission was warned by its own health specialists that radiation in the mines would cause an epidemic of lung cancer, but the AEC refused to warn the workers or eliminate the risk. The miners were warned only in the late 1960s.[1]

Such abuse constitutes genocide of an indigenous population. The 450,000 uranium workers from Oberrothenbach, Germany, are another huge exposed population, where meticulous medical records have been kept by the former East Germans. At least 20,000 have already died, and the records were only released in 1989 after the reunification of Germany. Analysis of this cohort will be as important as the Nagasaki/ Hiroshima data.[2]

At certain levels of the Australian government, the hazards of uranium mining—and of radon—are recognized, and stringent regulations have been passed to protect miners' health. Thus, mines must be kept dust-free, and miners are required to wash their hands and faces thoroughly before they eat, in order to remove radium-contaminated dust. These hygienic measures, however, are inadequate. The miners I spoke with said they had never been explicitly informed of the possible long-range, cancer-causing, and genetic effects of radiation; moreover, they reported that safety regulations are not always enforced.

In 1976, I visited the only Australian uranium mine then in operation, the Mary Kathleen Mine located in hot, dry Queensland. I was given a tour and later spoke to the miners about the medical and military dangers of uranium and uranium mining. I asked the plant manager whether radon alpha emission near the men was being measured; he replied that it was not. The only radiation monitored was gamma radiation from the uranium ore at the mine face. The radon was safe, he told me, because the mine was open cut, above ground, and "the wind blows the dust and gas away." His rationalization was patently feeble. As we stood there, talking in the open in the heat of the day, the air was completely still; dust hung heavily around the mine and swirled about the trucks as they moved along the roads. One man was totally enveloped in dust as he sat in the open cabin of his truck while a load of freshly quarried ore was dumped into his vehicle's tray.

After I spoke to the workers at this mine, several told me that management had recorded high levels of radiation in their urine. Concerned, now that they had been informed about the dangers associated with radiation, they asked me what this condition might imply. I explained that by the time radiation is detected in the urine, it has probably already been deposited in various organs of the body, where it may have done irreparable damage. Australia has since opened a huge uranium mine at Olympic Dam in South Australia and two others in the Northern Territory. She now exports uranium to nine countries to fuel their nuclear power plants, and the government is currently contemplating export to the nation of Indonesia.[3]

Australians have refused to contemplate nuclear power generation for themselves on the grounds that it is too medically dangerous. However, in June 1993, it was announced that a small research reactor at Lucas Heights on the periphery of Sydney, the only one in Australia, discharges more radioactive waste into the air and water than bigger, more powerful plants overseas. Apparently this reactor has discharged more radioactive iodine 131 into the air than the huge and dangerous nuclear fuel reprocessing and power plant at Sellafield in England. Radioactive water from the reactor is discharged into the sewer, and thence into our beautiful surfing ocean.[4]

MILLING

The U.S. Department of Energy (DOE) mills the uranium ore at taxpayers' expense. After uranium is mined, it is ground, crushed, and chemically treated. The end product is "yellowcake" (uranium 308), a uranium compound. The waste ore, called tailings, is discarded outside the mill and left lying on the ground in huge mounds; over the last forty years, more than 100 million tons have been accumulating in the American Southwest. These tailings contain the radioactive materials thorium (a gamma emitter with a half-life of 80,000 years) and radium.

Hundreds of acres of tailings thrown off from a once-thriving uranium business lay on the ground in Grand Junction, Colorado, until the mid 1960s, when city contractors hit upon the idea of using such tailings for cheap landfill and concrete mix. Construction firms used the waste ore to build a variety of structures including schools, hospitals, private homes, roadways, an airport, and a shopping mall. In 1970, local pediatricians noticed an increase of cleft palate, cleft lip, and other congenital defects among newborn babies in the area. Further investigation revealed that all these children had been born to parents living in homes built with tailings, and, when tested, many of these buildings demonstrated high radiation levels. The tailings emitted gamma X-ray-like radiation continuously, and also discharged alpha-emitting radon gas into the air—the same material that gave uranium miners lung cancer.

Soon after this discovery, medical professionals at the University of Colorado's medical center obtained funding from the U.S. Environmental Protection Agency (EPA) to study the possible correlation between low-level radiation and a rise in birth defects. But a year later, funds for the project were cut off; federal authorities claimed that the government had to cut back on many programs for budgetary reasons. Clearly, it did not consider the Grand Junction study worth pursuing. It is our moral responsibility to these—and all—children, however, to undertake such follow-up studies. A similar instance of laxity has occurred in Australia.

The town of Port Pirie, South Australia, once hosted a government uranium mill, which over the years generated some sixty acres of tailings, which were dumped in a nearby tidal dam. After the mill closed, children began using the dam area as a favorite spot to practice cricket, ride bikes, and play. During the late 1960s, a company specializing in rare earth materials bought the plant, refined pure thorium from imported rare earth sands, and exported it to Japan for use in color television sets. Soon after operations began, a shipload of the material was rejected at its Japanese port of entry when it was discovered that many barrels had ruptured during the rough voyage and were leaking the thorium onto the ship's deck. Sensitive to the hazards posed by radiation, and upset by the shipper's negligence, the Japanese refused the cargo and sent it back to Port Pirie—where it was immediately dumped into the tidal dam. Tha dam is periodically submerged by seawater, and the radiation was undoubtedly absorbed by algae, molluscs, fish, and other marine life. At low tide, children would play in the dam, crawling through the empty thorium barrels, or would fish in a nearby creek.

Radiation levels were monitored at the site in 1976, when the mill was again put up for sale, and in many locations they were found to exceed World Health Organization standards. This disturbing data was eventually leaked to the press, initiating a public outcry: parents feared their children might contract leukemia or other forms of cancer. Government officials claimed that the children were in no danger and that there was no need even to have them checked regularly for possible ill effects. The parents apparently accepted the officials' word, and no program for medical follow-up was instituted. The dam has since been covered with fill, but no doubt radium is still leaching into seawater and concentrating in the sea life.

ENRICHMENT

Typically, out of each ton of uranium ore extracted from the earth, four pounds is pure uranium. Of this amount, 99.3 percent is the unfissionable isotope uranium 238. Only somewhat less than half an ounce is the fissionable uranium 235. Since a specific minimal concentration of the latter isotope is needed for the fission process in most commercial reactors, but only 0.7 percent of the uranium found in its natural state is of this variety, uranium ore must be "enriched" so that its uranium 235 content comes to compromise approximately 3 percent of its bulk.

The enrichment process is extremely expensive, consumes vast amounts of energy (the Oak Ridge, Tennessee, enrichment plant consumes the electricity provided by two 1,000-megawatt reactors), and radioactive tailings, so-called depleted uranium, containing unusable uranium 238, are left lying on the ground. Because uranium 238 is an extremely dense material, it is used to make the cores of shells for conventional weapons, giving the weapons the capability of penetrating the armor plating of a tank. It is provided free to weapons manufacturers by the government-sponsored nuclear industry.

During the six-week land war against Iraq in 1991, at least 10,000 uranium 238 shells were used and at least 40 tons of this material was dispersed in Iraq and Kuwait. Children now play with empty shells; thus, they are exposed to external doses of gamma radiation, as well as inhaling and ingesting uranium particles, which can cause kidney disease, lung cancer, bone cancer, and leukemia. The death rate of children under five has doubled; in the first eight months after the war 50,000 children died of various causes; their diseases included cancer and stomach ailments. The United Kingdom Atomic Energy Authority reports that this 40 tons of uranium could cause tens of thousands of deaths and contaminate soil and drinking water in Iraq and Kuwait forever.[5]

Enrichment is so costly that in the United States the federal government has had to underwrite the process and operate America's three enrichment plants, located at Oak Ridge, Tennessee; Paducah, Kentucky; and Portsmouth, Ohio. Such subsidy has enabled America to become the world's chief supplier of enriched uranium. It has been a powerful position to hold, for it has given America almost complete control over, and responsibility for, the growth of nuclear power throughout the free world. The National Energy Bill passed by Congress in 1992 has renamed the uranium enrichment facilities as the U.S. Enrichment Corporation, to be a quasi-independent group, which will be operated with government assets; public shares, however, can and will be sold off in transition to privatization.[6]

The United States may soon lose its nuclear clout, however. Not only are its uranium reserves declining, but South Africa, France, Holland, the Confederation of Independent States (CIS), Britain, and Japan have already built enrichment plants of their own. Other countries almost certainly possessing uranium enrichment facilities are Iraq, Pakistan, India, Brazil, Argentina, Israel, and North Korea.[7] These foreign installations should be recognized as a legitimate cause for concern. They may well make access to nuclear fuel and the subsequent manufacture of nuclear weapons uncontrollable. Approximately two tons of weapons-grade uranium and plutonium disappeared from nuclear facilities in the United States during the 1970s. Some of it may have been stolen—possibly, as suggested by one CIA report, by Israel. Unfortunately, there is every indication that such acts, whether perpetrated by nations, or terrorist groups, or even criminal elements, will become a standard feature of a nuclear-powered world.

FUEL FABRICATION

After undergoing the process of enrichment, the uranium—in the form of uranium oxide—is converted into small pellets. These cylindrical pellets are then placed into 12–14-foot-long metal fuel rods less than one inch in diameter. A typical 1000-

megawatt reactor contains about 50,000 fuel rods comprising more than 100 tons of uranium in a cylindrical space about twenty feet in diameter and fourteen feet high. During the fabrication of these pellets, workers are exposed to the dangers of low-level gamma radiation emitted from the enriched fuel, and contamination by alpha emitters—radon, radium, and uranium.

NUCLEAR REACTORS

When the fuel rods are packed into the center (or "core") of a nuclear reactor and covered with water, the enriched uranium is ready to undergo fission. During this process, the nucleus of a uranium 235 atom breaks apart into fragments (or fission products—the nuclei of lighter atoms such as strontium or cesium), plus heat and one or more free neutrons. The neutrons released by the splintering of each nucleus in turn break up the nuclei of other atoms. When each free neutron absorbed by a uranium nucleus is replaced by a free neutron released by another fissioning atom, the reactor "goes critical" and the chain reaction becomes self-sustaining. Control rods, which absorb the fast-moving neutrons, regulate the speed of the process.

Fission releases a tremendous amount of heat, which is used by the reactor to boil water. The boiling water produces steam, which turns a turbine and generates electricity. Nuclear fission is thus a most dangerous and expensive method of boiling water—analogous to cutting butter with an electric saw.

In addition to generating electricity, uranium fission throws off hundreds of radioactive isotopes—all carcinogenic and mutagenic, with half-lives ranging from several seconds to seventeen million years or longer. The control rods, the cooling water, and the reactor vessel all become extremely radioactive over time, and the postfission irradiated fuel is one million times more radioactive than fresh fuel.

The operation of nuclear power plants presents many hazards that have been disregarded because, as frequently happens, short-term profit has outweighed the responsibility of industry and government to protect the public.

On February 2, 1976, three men with fifty-six years of combined work experience at all levels of the nuclear power industry resigned from secure (and well-paying) positions as nuclear engineers at General Electric. Dale Bridenbaugh, Richard Hubbard, and Greg Minor explained their reasons for leaving to the Joint Committee on Atomic Energy:

> When we first joined the General Electric Nuclear Energy Division, we were very excited about the idea of this new technology—atomic power—and the promise of a virtually limitless source of safe, clean and economic energy for this and future generations. But now . . . the promise is still unfulfilled. The nuclear industry has developed to become an industry of narrow specialists, each promoting and refining a fragment of the technology, with little comprehension of the total impact on our world system. . . . We (resigned) because we could no longer justify devoting our life energies to the continued development and expansion of nuclear fission power—a system we believe to be so dangerous that it now threatens the very existance of life on this planet.
>
> We could no longer rationalize the fact that our daily labour would result in a radioactive legacy for our children and grandchildren for hundreds of thousands of years. We could no longer resolve our continued participation in an industry which will depend upon the production of vast amounts of plutonium, a material known to cause cancer

and produce genetic defects, and which facilitates the continued proliferation of atomic weapons throughout the world.

During their long testimony, these men claimed, among other things, that the defects and deficiences in just the design of nuclear reactors alone created severe safety hazards, and that the combined deficiences "in the design, construction, and operation of nuclear power plants makes a nuclear power plant accident, in our opinion, a certain event. The only question is when and where." Since that time, Three Mile Island and Chernobyl have melted down, with the strong possibility of more meltdowns to come.

What makes an accident in a nuclear power station uniquely dangerous is the potential release into the environment of highly poisonous radioactive elements that can contaminate large areas of land and make them uninhabitable for thousands of years. What makes an accident seem inevitable is the human factor. The most advanced plant is still at the mercy of the fallible human beings who design, build, and operate it. Millions of parts are needed to construct a nuclear reactor, and each must be made, assembled, and operated with little room for error. Remember also that the controlled chain reaction in a reactor is just a silent bomb waiting to melt down or to spread its toxic waste within the natural environmental systems for the rest of time.

The design of a nuclear power plant is extremely complex; its construction is very difficult. As the resigning engineers noted, these aspects of plant management are in the hands of specialists who do not necessarily understand the work of—or even communicate with—other specialty groups. In fact, today no one individual or group coordinates the complete process of building and operating such a reactor. The U.S. Nuclear Regulatory Commission (NRC), formerly the Atomic Energy Commission—with no appreciable change of staff—is in charge of licensing and inspecting each plant. Plants are fully inspected before operation, but inspection thereafter is often spotty. All U.S. nuclear power plants now have resident NRC inspectors in the wake of the Three Mile Island catastrophe, but daily responsibility for plant inspection and radiation monitoring still rests with the licensees.

Nuclear power plants operate under many untested theoretical principles. Certain safety systems are built on the shaky test results of computer models. Many components are made of metals susceptible to failure from contact with the nuclear environment. As a result, corrosion causes cracks and subsequent leaks that are often difficult to remedy in certain sections of the plant, because localized intense radioactivity prevents entrance. Nuclear plants are built to leak, even when functioning perfectly. Routine emissions of radiation are discharged daily into the air and water. The nuclear industry has avoided making certain necessary repairs by conducting computer studies that prove that a particular pipe or component part was unnecessary in the first place. In engineering language, this practice is termed a "fix." In April 1993, the NRC identified fifteen reactors in the U.S. whose metal reactor vessels are so weakened by radiation that they are probably unsafe. The reactors are aging faster than their designers anticipated.[8]

In a nuclear plant, the repair of a mere pipe, a simple task under ordinary conditions, often requires that the plant be shut down; many workers must be called in. Some years ago, a pipe failed at the Indian Point 1 reactor, located on the Hudson River forty miles north of New York City. As a result, the plant was rendered inoperable for six months and 1,700 certified welders—almost every certified welder in the Consolidated Edison company—were needed to repair the damage. It was necessary to hire so many because within a few minutes each worker would receive the dose of radiation deemed allowable for a three- to six-month period.

Like the workers at the Mary Kathleen uranium mine, workers at nuclear power plants in America are told in only general terms that radiation is dangerous; they are not informed specifically of its carcinogenic and mutagenic properties. While each person must wear a badge that monitors the level of radiation to which he or she is exposed, this device registers only gamma radiation and perhaps a small quantity of beta; it does not measure the internally emitted radiation—the alpha or beta emitters that can be swallowed or inhaled. Furthermore, workers are permitted to receive thirty to a hundred times as much radiation per year as the limits set for the general public.

At present the nuclear industry is required to keep employee medical records for no more than five years while workers are on the job. They can be destroyed when employees leave. Radiation records must be kept, but detection monitors do not record doses below a certain level, and it is difficult to assess the relationship of any health problems to radiation. Medical follow-ups are not conducted to determine whether former employees have contracted malignancies, or to check whether their children have been genetically affected by preconception irradiation to their father's or mother's gonads, or by exposure of the mother during pregnancy. (In the event of such disability, no compensation is paid.)

At many plants the hiring of unskilled or itinerant migrant labor is a common practice. These employees work briefly for high wages, often in areas of intense radiation. They are euphemistically called "sponges" in the nuclear industry. After receiving their legal maximum doses at one facility (sometimes in only one day or less), they may be hired at another plant, often without being questioned concerning their previous radiation exposure. Plants hire these transients in order to preserve the "body banks" of their full-time employees. At a reprocessing plant in West Valley, New York, which is now inoperable, "fresh bodies" were often recruited from local colleges and

months at a time. Continual brownouts and
t so try the public's patience that safety con-
sacrificed. A really serious accident could lose
lear generating capacity. But remember that
ould eliminate the need for nuclear power. If all
ing their washing in the sun outside or beside
the winter and abandoned drying machines,
would not be necessary.

first nonmilitary reactor began operating in
7 and was soon plagued with breakdowns and
he late sixties, nuclear plants in Britain, the U.S.,
Switzerland had all experienced major accidents
esulted in death or injury to personnel and were
gerous to the general public.

accident at Britain's Windscale reactor on the
sed a cloud of radioactivity across neighboring
ges. Soon afterward, the government supervised
thousands of gallons of contaminated milk into
This action may have spared the public tempo-
for long. The polluted ocean water contami-
life and fish-eating mammals, and the radioac-
lly worked its way up the food chain. Later
d high levels of iodine 131 in the thyroid glands
nts as a result of inhalation of airborn iodine 131
of contaminated food.

he Windscale accident, the British government
local farmers for the contamination and loss of
similar major accident or meltdown occurred in
ates, however, few people would be adequately
or radiation-induced illness, or loss of life or
ause the liability of the U.S. nuclear industry is
1950s, private insurance companies recognized
ower generation was an enterprise fraught with
dicted that a nuclear accident of substantial pro-

bars to do the industry's dirty work. From a medical stand-
point, this practice is unethical. Furthermore, it is illegal to
overexpose any worker. It still happens.

The nuclear power accident that poses the greatest threat to
public safety is termed a "meltdown" or the "melt through-to-
China syndrome." Such an event almost occurred at Three
Mile Island, where in fact the fuel partially melted. Seven years
later at Chernobyl, most of the fuel did melt, and the molten
material disappeared into the bowels of the plant, where it
remains today. The reaction created a massive hydrogen explosion, which released one-third of the radioactive inventory.
These accidents can be initiated by a pipe breakage, or safety
failure, or human error, which either permits the cooling water
at a reactor's core to drop below the level of the fuel rods or
removes the minimum number of control rods. The fuel rods
then become so hot that they melt; then the whole mass of
molten uranium burns through the reactor vessel and the
"container" (the concrete base of the plant) into the earth,
triggering a steam explosion that blows the containment vessel
apart, releasing its deadly radioactive contents into the atmosphere.

Soon after a severe meltdown, thousands of people will die
from immediate radiation exposure; and more will perish
weeks later of acute radiation illness—as they did at Chernobyl. Food, water, and air will be so grossly contaminated, as at
Chernobyl and its environs, that in five years there will be an
epidemic of leukemia and immune system disfunction leading
to an increase in a spectrum of disorders. Then fifteen to sixty
years later an increase in solid cancers can be expected. The
genetic deformities that might appear in future generations are
hard to predict, but they will surely occur.

Such a meltdown was examined in the 1970s by the Union
of Concerned Scientists (UCS), who conducted a two-year
study of a hypothetical "expanded nuclear economy," as

planned by the nuclear industry. They concluded that before the year 2000, close to fifteen thousand people in the United States may die of minor nuclear accidents; moreover, the UCS also estimated that in the same period there was a one percent chance that a major nuclear accident would occur, killing nearly 100,000 people; most of whom would die of radiation-induced cancers. Even the U.S. Nuclear Regulatory Commission (NRC) itself calculated the risk of a major nuclear accident to be 45 percent between the years 1985 to 2005; and yet they didn't close down the industry.[9] Although somewhat delayed, an "expanded nuclear economy" is now being planned by the nuclear industry.

In addition to Three Mile Island, when the United States came frighteningly close to experiencing a catastrophic meltdown, in 1975 the Brown's Ferry atomic power plant near Athens, Alabama, witnessed the worst accident to date of the U.S. nuclear industry. Two electricians, using candles to check for air leaks, accidentally set fire to some highly combustible polyurethane foam that was being used as a sealant. The fire quickly spread to the plastic cables surrounding the cables controlling the operation of the reactor and the emergency core cooling system (ECCS)—a system designed to prevent a meltdown by piping in more water if the coolant water in the reactor begins to fall uncontrollably below safe levels. The blaze raged in the bowels of the plant for seven and a half hours. In the process it severed thousands of cables, knocking out most of the reactor's control systems and the ECCS. Operators watched helplessly as the water level in the reactor core dropped sharply. At last, after workers resorted to equipment not intended for emergency cooling services, the water began to rise. A major disaster was thereby averted, but by accident, not by design.

A wide array of regulatory violations and inadequate safeguards were disclosed in the wake of this accident. First, it was

revealed that the N
only if a plant were
company monitor
was not thus insure
inspections of any
the cables were loc
a single mishap wo
trolling the reactor'
Third, although the
zens were never ale
into effect—becaus
This practice has si
documented review
tors called the fire h
tions by the NRC r
resident inspectors.[

The fact that son
same design had bee
of the nuclear indust
are unique designs.)
was found to be at
should also be shut c
now the NRC will u
exists at a plant simi
plants and inform th
their responses, the 1

Since the industry'
half their time idle bec
involving safety-relate
improve, so that duri
reactors was in the mi
contributes to 10 perc
consumption.[12] Henc
sections of the nation

breakdowns fo
blackouts migh
cerns might be
the whole nuc
conservation w
U.S. citizens h
the furnace in
nuclear power

The world'
Canada in 194
accidents. By t
the USSR, and
as well, which
potentially dar

In 1957, an
Irish Sea relea
fields and villa
the disposal o
the sea nearby
rarily, but no
nated marine
tivity eventua
studies reveale
of local reside
and ingestion

Following
compensated
their milk. If
the United S
reimbursed f
property, bec
limited. In th
that nuclear
risk: they pre

portions could financially destroy them, and they were therefore reluctant to provide insurance. Because Congress was eager to promote nuclear power, however, it passed the Price Anderson Act, which absolved America's power companies of major financial responsibility in the event of a nuclear disaster. This act recognizes that the nuclear industry could never have developed if private insurance had had to cover all contingent liabilities.

The Price Anderson Act limited U.S. government and private corporate liability to $560 million for each nuclear generator until 1988, when Price Anderson Act amendments raised the limit to $7.8 billion per accident. The utilities now cover a large fraction of this insurance themselves.[13] Obviously, a major nuclear accident would cost billions of dollars in property damage alone, not to speak of massive loss of life. What dollar value can we attach to each life lost?

The fact that the Three Mile Island accident has resulted in little compensation means that no epidemiological studies have been conducted on this irradiated population to determine the incidence of radiation-induced disease, although a recent claim was covered for a woman who delivered a child with Down's syndrome after the accident.

In the event of a nuclear mishap, individual citizens have no recourse to their own insurance firms. Most homeowners' policies include a "nuclear exclusion" clause, which denies an individual compensation for a "nuclear catastrophe"— whether caused by nuclear war or nuclear power generation.

While a major nuclear accident might result from internal plant safety failures, nuclear reactors are also vulnerable to terrorist takeovers and internal sabotage by dissatisfied or simply unstable employees. A U.S. government study conducted in the 1970s concluded—in all seriousness—that, in order to ensure adequate protection, every plant should: 1) employ a full-time guard armed with a bazooka, to shoot down any threatening helicopters and aircraft; 2) obtain the services of a psychiatrist, to assess the behavior of employees; and 3) discourage gambling amongst workers lest it attract loan sharks and mobsters. In fact, on February 7, 1993, a former mental patient drove his station wagon past the guard at the remaining operating Three Mile Island nuclear reactor, crashed it through a metal door, and drove sixty feet inside the turbine hall where he remained for four hours. Several weeks later, a letter related to the February 26th bombing of the World Trade Center referred to future "Nuclear Targets."[14]

In fact, when a country is dotted with nuclear power plants, enemies, terrorist organizations, or mentally deranged people need no nuclear weapons to wage nuclear war; they need only drop a conventional bomb on a nuclear reactor to release most of the radiation it contains, killing thousands of people. It is obvious that if Europe had been powered by nuclear generators during the World War II, it would still be uninhabitable due to widespread radioactive contamination of air, food, and water.

The dangers intrinsic to the nuclear industry are unique. Not only is nuclear technology unsafe, the industry has virtually ignored the fact that a reactor's highly radioactive waste products are indestructible. Were all the dangers I have enumerated thus far miraculously overcome, the sole problem of the impossibility of waste "disposal"—and its impact on future generations—should be sufficient to give pause. For, once created, some of these nuclear by-products will remain in our biosphere for tens and hundreds of thousands of years, wreaking irreversible damage on plant, animal, and human life, unless successfully contained. What moral right do we have to leave such a legacy to our descendants?

Chapter 4

Nuclear Sewage

The only commercial nuclear fuel reprocessing facility ever to operate in the United States was located in West Valley, New York, about thirty miles south of Buffalo. From 1966 (the year that it opened its doors) through 1972, the plant repeatedly violated radiation exposure standards. In October 1976, Nuclear Fuel Services, Inc., the private firm that owned and operated the facility, closed the plant permanently—bequeathing tons of radioactive waste to the citizens of New York State.

In 1977, the U.S. Congress's Committee on Government Operations spent $1 million to determine what might be done with West Valley's 600,000 gallons of high-level, neutralized liquid waste and 2 million cubic feet of low-level solid waste. It concluded at that time that the problem was "gargantuan" and might cost as much as $600 to $700 million to remedy.

While in Washington, D.C., to address the 1977 Convention of the Association for the Advancement of Science, I was

invited to the office of the congressman who represented the
West Valley district. To my surprise, he proceeded to ask me
what I thought could be done. I replied that I did not know,
for as a physician, I have no answers for the problems posed
by the "back end" of the nuclear fuel cycle: the safe storage of
the lethal radioactive wastes produced each year by America's
commercial and military nuclear programs. Nor, unfortunately,
does anyone else—including the nation's nuclear engineers.

The term *nuclear waste (radwaste)* refers to all the unusable,
radioactively contaminated by-products of the nuclear fuel
cycle and the weapons program. Intensely radioactive, "high-
level" waste consists of fission products of uranium in the
form of either intensely radioactive irradiated fuel or as a
concentrated liquid or solid. Both forms contain substantial
quantities of deadly plutonium. "Low-level" waste includes
contaminated articles of clothing, decommissioned plant components, and fission by-products given off by nuclear reactors
in dilute aqueous and gaseous form. But often so-called low-
level waste is highly radioactive.

Because hundreds of radioactive elements are produced
when uranium is consumed in the fission process, they gradually build up in the fuel rods and begin to hinder its efficiency.
In addition, some uranium 235 remains unfissioned, and it,
together with the plutonium, can be retrieved and reprocessed
for use at the front end of the fuel cycle, or to be manufactured
into nuclear weapons. Because of the inefficiency factor, each
year reactors are shut down so that technicians can replace
one-quarter to one-third of the contaminated fuel rods.

Since there is no way to dispose of the highly radioactive
irradiated fuel rods permanently, they are currently managed
on site at each of the operating and formally operational reactor sites in the U.S. They must be handled with extreme care
because they are intensely radioactive; a few seconds of un-
shielded exposure would deliver a lethal dose.

The rods are also extremely hot and must be stored for five years or more in a pool of water usually located near the reactor.[1] The water cools the rods, preventing them from spontaneously melting and releasing their poisonous contents into the atmosphere, and permitting their radioactivity to decline. The swimming-pool-like structures are designed to hold the spent fuel for one to ten years. These temporary storage pools often leak radioactive water into the environment. The water has to be filtered and the resultant sludges comprise some of the hottest "low-level" radioactive waste. In the nuclear weapons industry, the rods have been sent to a reprocessing plant in order to extract the residual uranium 235 and plutonium 239.

In 1977, President Carter declared a moratorium on the operation of commercial reprocessing plants in the United States, in the hope that other countries would follow this nation's example and refrain from reprocessing their irradiated fuel to build atomic bombs. Reagan rescinded this moratorium shortly after he took office.

Because there are no commercial reprocessing plants in operation in the United States today, individual nuclear plants are required to store their irradiated fuel in the holding pools (euphemistically called swimming pools) indefinitely, and the accumulation of such rods is causing the pools to become dangerously full. The NRC has granted nuclear plants an interim license to pack their fuel rods closer together than originally planned,[2] a procedure that is only relatively safe if done correctly and operated properly.

In certain circumstances, accidents in these "swimming pools" could become very severe, such as the loss of cooling water, with subsequent overheating of the zirconium cladding of the rods. This would cause a steam zirconium reaction, with rapid dispersal of the toxic fission products into a building not designed for their containment, and certain systems that were designed to prevent such an accident could fail or be rendered inoperable.

Criticality in the pool is also a possibility. By the end of 1976, 3,000 metric tons of spent fuel lay in nuclear pools across the United States. By 1983, an estimated 13,000 tons was dangerously straining the capacity of storage pool facilities, and in December 1989, 16,423 metric tons of spent fuel had accumulated at commercial reactor sites.[3]

The U.S. military reprocessed the fuel from its nuclear reactors by first dissolving the irradiated fuel rods in concentrated nitric acid. The solution that results is highly corrosive. The volume ratio of fuel rod to reprocessing liquid is one to 200,000. That is 200,000 times more volume for the same amount of radioactivity.

Today there are more than 92 million gallons of this high-level liquid waste in storage tanks in the United States; many of these liquids are so hot that they boil spontaneously and continually. Most can be found at the Hanford Military Reservation in Washington and the Savannah River facility in South Carolina. The major portion of these wastes lie in huge carbon steel tanks that cannot withstand the waste's corrosive properties for more than twenty-five years; newer stainless steel tanks can last for fifty. But what is half a century compared to the thousands of years that this radioactive material must be kept isolated from the environment?

Moreover, many of these tanks have already sprung deadly leaks. From 1958 to 1975, twenty of Hanford's older, single-walled carbon steel tanks developed cracks through which 430,000 gallons of high-level waste leaked into the soil. In 1973 alone, an oversight on the part of a tank operator caused the escape of 115,000 gallons of high-level radwaste into the environment. The Hanford Reservation is located several hundred feet above the Columbia River system, which supplies drinking water to cities in the Northwest.

Eight double-walled tanks at the Savannah River facility have shown stress corrosion cracks. In one tank, 175 cracks were detected; and 100 gallons of high-level waste leaked into the soil adjacent to the Savannah River. Local rainfall—four inches per month—may hasten migration of the isotopes into the river.

Between 1946 and 1970, the U.S. military encased 47,000, 55-gallon drums of low-level waste in concrete-lined steel drums and dumped them into the Pacific Ocean about thirty miles outside the San Francisco Bay. Over one third of these drums have now leaked radioactivity into the Bay area's major fishing grounds. Giant sponges, many of them over three feet high and believed by some to be mutants, have attached themselves to the drums.

In an attempt to minimize the corrosive properties of West Valley's high-level waste, scientists used sodium hydroxide to neutralize the acid solution in which the irradiated fuel rods were dissolved. Unfortunately, this procedure results in greater problems than it solves, for it doubles the waste volume and, over a period of time, precipitates a radioactive sludge consisting of strontium 90, cesium 137, other deadly fission products, and some plutonium to the bottom of the tanks. Because of the probable high concentration of plutonium in this sludge, some experts fear that it may go "critical," initiating a reaction similar to a meltdown and releasing tons of deadly radioactive materials into the biosphere. If the 600,000 gallons of high-level waste stored at West Valley were to be dispersed in this way, the resulting radiation could devastate Buffalo and its surrounding towns and land.

The consequences of such a disaster were foreshadowed by an accident that occurred in 1957 in Russia, at a vast nuclear complex in Kyshtym, a small town in the Ural Mountains. According to the Soviet scientist Dr. Zhored Medvedev, an "enormous explosion, like a volcano of radioactive waste," dispersed clouds of radioactivity "over hundreds of miles." He reports, "Tens of thousands of people were affected, hundreds dying, though the real figures have never been made public." The area around Kyshtym is now a wasteland devoid of life. It will remain uninhabitable virtually forever.

This Soviet disaster, apparently at a waste repository site, was revealed in CIA reports that were released in 1977 under the Freedom of Information Act. It is not unfair to ask why the U.S. government covered up the evidence of this accident for almost twenty years.

But in 1989, Moscow revealed that 20 million curies of long-lived isotopes were released over a 410-square-mile area, contaminating over 250,000 people. Most persons in the area have lost a friend or relative from cancer. Radiation levels still measure eighty-five times normal background.

Despite this disaster, production of plutonium continued at the site until 1990, and reprocessing goes on. The nearby village of Metlino was evacuated in 1952, and the village became a radioactive waste lake, called Reservoir No. 9, or Lake Karachay, as the nuclear production facility continued to dump their nuclear waste into the lake. Radiation levels here measure 250 times normal.

Another massive accident at Lake Karachay in 1967 exposed some 50,000 people to radioactive fall out. A hot summer was followed by a dry winter, and dust from the lake was blown over an 1,800-square kilometer area, contaminating 41,-000 people, with high levels of radioactive isotopes. Many of these same people had been previously irradiated in the 1957 Kyshtym disaster.[4]

From 1949 to 1963, the testing of nuclear weapons at Semipalatinsk in Kazakstan exposed over 1.5 million people to radiation. In fact, officials in the CIS now admit that millions have been injured or have died because of radioactive fall out.[5]

(For further reading of the former Soviet Union's disastrous nuclear weapons complex and subsequent health effects, I refer you to the excellent paper, "Russian/Soviet Nuclear Warhead Production," by Thomas B. Cochran and Robert Standish Norris.[6])

In the U.S., many low-level wastes accumulate during the daily operation of nuclear reactors. Solid wastes in this category (contaminated plant equipment and clothing) were buried in a totally irresponsible fashion straight into the ground with no containment, no monitoring, and no plans for excavation and subsequent retrieval, at Sheffield, Illinois; Maxey Flats, Kentucky; Barnwell, South Carolina; Rocky Flats, Colorado; and Hanford, Washington.

Although records of buried plutonium waste are incomplete at the time of writing, it is estimated that 1,696 pounds of plutonium were buried at six shallow land sites in the U.S.— Hanford, Idaho National Engineering Laboratory, Lawrence Livermore and Sandia Labs, Oak Ridge National Laboratory, and the Savannah River site. In 1988 it was estimated by the Department of Energy Integrated Data Base that this amount would multiply 3.5 times by the year 2013.

A total of thirty-five tons of plutonium is contained in nuclear waste throughout the world.[7] The many and varied radioactive elements emanating from these wastes are often leached into the soil by rainfall. In addition, these materials are readily susceptible to disinterment: several years ago, it was discovered that contaminated gloves, wheelbarrows, and shovels were missing from an unguarded, unmarked burial site in Nevada; it was later revealed that local citizens, totally unaware of the dangers involved, had dug up and used the equipment.

Animals also act as carriers of radioactive material. A study at the Hanford Reservation showed, for example, that jackrabbits had spread radiation over a wide area. They picked up the material by burrowing near or into trenches where radioactive material had been buried. They obviously ate some of this material, since traces of the radioactive isotopes were found in their feces. Similar amounts of radioactivity were also found in the feces of coyotes and the bones of dead hawks—animals that had apparently eaten the radioactive jackrabbits. Such studies are few and far between, but this sort of activity must be ubiquitous at radioactive waste dumps.

Government regulatory policies permit a certain amount of low-level liquid waste to be released to burden the environment. Nuclear advocates and government regulatory agencies claim that this liquid effluent leaks into the environment in "safe concentrations" and that "routine emissions" of this type containing long-lived isotopes are even further diluted by the ecosphere. However, this is aberrant reasoning, because even if these isotopes are very dilute, they will inevitably become reconcentrated in the aquatic food chain.

Government agencies similarly permit a certain amount of radiation to be "routinely" emitted from a reactor stack, arguing that the gaseous radioactive effluent consists primarily of the noble gases krypton and Xenon, two elements that will not be incorporated into biological systems to any appreciable degree for "many years." Xenon 135, however, decays into cesium 135 with a half-life of 3 million years.[8] Nevertheless, scientists are uncertain about the long-term effect of krypton emission on the environment. A gamma emitter, krypton has been found to concentrate in the fat layers of the lower abdomen and upper thighs, near the gonads. To claim that the radioactivity in these discharges is diluted to safe levels is fallacious: it in fact adds to the existing levels of background radiation, increasing the risk of disease.

In addition to "routine emissions," "abnormal releases" of radiation often occur as a result of plant accidents. Frequently reported to the press, these releases are usually accompanied by a statement from the NRC claiming that the amount of radiation that escaped was lower than the normal background radiation (the implication is that it was therefore "safe"). Missing from these accounts is the fact that any release of radiation adds to the level of background radiation, thereby increasing the risk of radiation-induced cancers and mutations.

Among the many examples of "abnormal releases" are iodine 129; with a half-life of 17 million years, it enters the food chain and eventually concentrates in the human thyroid gland. According to a report issued by the U.S. Environmental Protection Agency in 1972, the iodine 129 removal system installed at the West Valley reprocessing facility, a private operation that reprocessed commercial fuel, did not perform according to design. As a result, during the plant's operation (from 1966 to 1971), 45 percent of the total content of iodine 129 in the reprocessed fuel escaped into the nearby environment. At a distance of five to six miles from the plant, specific activity levels of iodine 129 were found to be 10,000 times greater than normal background radiation; ten miles from the plant the levels were ten times higher.

Indeed, many such accidents led up to the disaster at Three Mile Island in March 1979. In 1976, Vermont Yankee in Western Massachusetts emptied 83,000 gallons of low-level waste water into the Connecticut River. In Florida, a leaking storage pool of irradiated fuel rod coolant dumped thousands of gallons into the ocean, because there was nowhere the radioactive fuel could be shipped to allow repairs to be made; and an explosion at Connecticut's Millstone I plant released excessive amounts of radiation into the environment. These are just a few of the events in the U.S. since the commercialization of nuclear energy.

Every nuclear power plant will eventually end up on the radioactive garbage heap, because a plant can operate for only twenty to forty years even when things go well, before it becomes too radioactive to repair or maintain. When the time comes for a plant's demise, it must be shut down and "decommissioned." It must be either disassembled by remote control (because it is simply too radioactive to handle manually) and its constituent parts buried, or the entire plant could be buried under tons of earth or concrete to become a radioactive mausoleum for hundreds of thousands of years. In either case, the remains must be guarded virtually forever. Alternatively, it is probably wiser to leave the reactor above ground, where it can be continuously monitored and removable if necessary and where it will remain a constant ugly reminder of the obsolete stupid and dangerous folly of the nuclear age.

A nuclear reactor constructed near Los Angeles underwent the delicate process of disassembly by remote control. Built in the 1950s, this small plant cost $13 million. The price of demolition in 1979 was estimated at $6 million. Larger reactors (1000-megawatt or greater) may in fact be impossible to decommission. Commonwealth Edison's Dresden I plant in Morris, Illinois, a 200-megawatt reactor that went into service in 1960, was permanently removed from operating service in 1978. At eighteen years old, the oldest commercial reactor in the U.S. had become too radioactive to repair. Dresden units II and III, both 800-megawatts, are still in service.[9]

Can we do anything to protect ourselves and future generations from the lethal legacy of nuclear sewage? At present, the answer is no. Technologists have offered a number of ingenious proposals, ranging from vitrification of high-level waste, i.e., converting it into glass, and burial in salt formations, to lowering waste into ocean trenches, burying it under Antarctic ice, or launching rockets loaded with radioactive waste into the sun. None of these techniques have been proven to be practical or safe.

Succumbing to technological fervor, the U.S. government prematurely committed enormous economic resources, together with political and scientific reputations, to a half-baked, power-generating technology that is neither clean, cheap, nor safe. The nation's public utilities should never have been permitted to proceed with nuclear energy production until they demonstrated that the public's health could be protected from the carcinogenic and mutagenic effects of its radioactive wastes. This was not done.

Industry engineers and physicists concede that the nuclear waste problem still remains to be solved, but in their public pronouncements they urge us to trust them, to have faith in their abilities and in the inevitable advance of technology. I have no confidence in this line of reasoning. It's as if I were to reassure a patient suffering from terminal cancer by saying, "Don't worry, my medical training will enable me to discover a cure."

Nor can technology alone ever provide the answers we seek. For even if unbreakable, corrosion-resistant containers could be designed, any storage site would need to be kept under constant surveillance by incorruptible guards, administered by moral politicians living in a stable warless society, and left undisturbed by earthquakes, natural disasters, or other acts of God for no less than half a million years—a tall order, which science cannot fill. I will discuss this in more detail in the chapter on Waste Cleanup.

Chapter 5

Plutonium

Plutonium is one of the most carcinogenic substances known. Named after Pluto, god of the underworld, it is so toxic that less than one-millionth of a gram (an invisible particle) is a carcinogenic dose.

One pound, if uniformly distributed, could hypothetically induce lung cancer in every person on earth.*

Found in nature only in a remote region of Africa, and in minute amounts, plutonium is produced in a nuclear reactor from uranium 238 in quantities of four hundred to five hundred pounds annually! This alpha emitter has a half-life of 24,400 years and, once created, remains poisonous for at least a half a million years. Plutonium may be more dangerous than

---

*Toxicity figures are for plutonium 239, the isotopes used to fuel atomic bombs. Other isotopes are produced in nuclear reactors, including plutonium (Pu) 238, Pu 240, and Pu 241. The mixture is 5.4 times more toxic than Pu 239).

originally thought. Irradiation of mouse and hamster cells by plutonium alpha particles creates chromosomal abnormalities that appear only after several generations of cell divisions. If this work is confirmed, the toxicity of plutonium may have to be revised upwards. It also calls into question what already lies ahead of us genetically as a species, due to our exposure to date.[1]

Plutonium is a chemically reactive metal, which, if exposed to air, ignites spontaneously to produce respirable-sized particles of plutonium dioxide, a compound also produced as a talcum-fine powder during fuel reprocessing. These particles can be transported by atmospheric currents and inhaled by people and animals. When lodged within the tiny airways of the lung, plutonium particles bombard surrounding tissue with alpha radiation. Smaller particles may break away from the larger aggregates of the compound to be absorbed through the lung and enter the bloodstream. Because plutonium has properties similar to those of iron, it is combined with the iron-transporting proteins in the blood and conveyed to iron-storage cells in the liver and bone marrow. Here, too, it irradiates nearby cells, inducing liver and bone cancer, and leukemia.

Plutonium's iron-like properties also permit the element to cross the highly selective placental barrier and reach the developing fetus, possibly causing teratogenic damage and subsequent gross deformities in the newborn infant. Plutonium is also concentrated in the testicles and ovaries where it inevitably will cause genetic mutation to be passed on to future generations and, in some cases, cancer of the testicles.

In 1969, the second largest industrial fire in U.S. history consumed two tons of plutonium at a military reactor site in Rocky Flats, Colorado. Miraculously, the fire was contained within the plant, but thirty to forty-four pounds of respirable plutonium escaped and contaminated parts of Denver, sixteen miles away. Radiation tests in the region demonstrated that thousands of acres of land, including a major water source, were contaminated. Off-site, some contamination resulted between 1959 to 1969 when plutonium-infected machine oil leaked from 3,000 barrels stored near the plant gates. Hearings held by officials of the Colorado Department of Health in 1973 revealed that many local farm animals were being born with grotesque deformities.

The food chain concentrates plutonium many times over, most commonly in fish, birds, eggs, and milk. However, since plutonium molecules are large, they are not usually absorbed directly into the body through the gastrointestinal tract, except by infants during the first four weeks of life (at which time their immature intestinal walls permit absorption). The extreme susceptibility of infants is compounded by the fact that plutonium concentrates in milk, whether from animals or humans. Chlorinated water enhances the absorption of plutonium through the gastrointestinal tract.[2]

Plutonium does not simply vanish at the death of a contaminated organism. If, for example, someone were to die of a lung cancer induced by plutonium, and were then cremated, contaminated smoke might carry plutonium particles into someone else's lungs. When a contaminated animal dies, its polluted carcass may be eaten by other animals, or its poisoned dust may be scattered by the winds, to be inhaled by other creatures. Similarly, the same plutonium particles can go on concentrating in the testicles and ovaries of successive generations of human beings or animals, conceivably causing repeated genetic damage for up to 500,000 years, while the damaged genes are themselves passed on from generation to generation.

Plutonium 241 has another deadly characteristic: it produces Americium, a by-product with a half-life of 460 years. Americium is even more potent than its parent: because it is more soluble than plutonium, it is more readily absorbed into the food chain. In recent years, this deadly material has been used to power tens of millions of ionizing smoke detectors, which, if damaged by fire, release this carcinogenic element into the air in powdered form, to be inhaled by unsuspecting people.

What will happen to these devices? Most often, they end up in local dumps, where the Americium eventually migrates into the soil and future food supplies. Many states, including California, require smoke detectors in hotels, rental rooms, and apartments. There is no program that requires collection of these extremely dangerous devices.⁵ The NRC has specifically exempted Americium 241 from any further regulations, so that it is free to be scattered to the four winds.

I testified in New York City when the NRC was considering using Americium in fire detectors in the late seventies. My discussion was factual and warned of the impending dangers. But they went ahead and used it anyway.

When, in 1941, the American physicists Edwin McMillan and Glenn Seaborg first identified plutonium 239, they also discovered that it was fissionable; in other words, it was raw material for atomic bombs. Four years later the U.S. military tested plutonium's fission power on Nagasaki.

Only ten to twenty pounds of plutonium are required to make an atomic bomb. Since each 1,000-megawatt reactor produces four hundred to five hundred pounds of the element per year, any nation possessing a reactor could hypothetically make twenty to forty bombs annually. Thus, even a small experimental facility can become an effective bomb factory. India proved this fact in 1974, when it used plutonium extracted from an experimental reactor bought from Canada to build and detonate the subcontinent's first homemade nuclear device.

In the 1970s, researchers at America's Oak Ridge National Laboratory designed a simple reprocessing plant that would take four to six months to build from readily obtainable, inexpensive materials. The first 22 pounds of plutonium 239, reprocessed from irradiated fuel derived from nuclear power reactors, would be ready within one week. The plant would produce 220 pounds of plutonium metal per month—enough for about twenty atomic bombs.

Plutonium's role in atomic bomb production has made its value soar on the black market. It is vulnerable to theft by non-nuclear nations, terrorists, racketeers, and lunatics. Once an individual or group is in possession of plutonium, bomb fabrication is not very difficult. Using only declassified information, college students have succeeded in designing functional bombs. The designs call for metal fixtures bought at local hardware stores and ten to twenty pounds of plutonium, an amount that can easily be concealed in a shopping bag.

Nuclear technologists look to plutonium as an eventual substitute for uranium 235 in nuclear reactor energy production although there is now a worldwide glut of uranium. To eliminate the need for it, the industry has designed what is known as the "fast breeder reactor," which is fueled by a combination of plutonium and uranium 238. It is called a "breeder" because in the process of generating electricity, it creates more plutonium than it consumes. The average estimated doubling rate is thirty to fifty years. In 1979, France's breeder reactor, the Super-Phoenix, was expected to take sixty years to duplicate its plutonium load.

However, in June 1992, the French government indefinitely postponed operations of the Super-Phoenix because they were worried about a nuclear meltdown and/or explosion, and the lack of adequate means to handle a liquid sodium fire.

Furthermore, Germany closed its fast breeder reactor in 1991, and abandoned its reprocessing plant. Britain plans to close its fast breeder prototype in 1994.[4] But the U.S. Department of Energy, under George Bush, was planning to construct a new generation of breeder reactors over the next thirty years.[5] President Clinton has since cut all funds for the development of these breeders, but private funding from the nuclear industry, or a new Republican president in 1996, could regenerate these plans.

However, Japan is about to embark upon a most ambitious and dangerous breeder program. Conceived in the early 1970s, when breeder reactors were assumed to be commercially viable, it is now proceeding apace.

Because Japan's reprocessing plant will not be completed until the late 1990s, it decided to ship its commercial spent fuel rods to England and France for reprocessing, with the express permission of the United States. (The original fuel rods were and are supplied by the United States and are therefore under U.S. jurisdiction.) The first shipment of 1.7 tons of separated plutonium is packed into a ship on the high seas as I write, traveling between the French port of Cherbourg and Tokyo. The route has been kept secret by the Japanese, a plan condoned by the U.S. government, which did not mandate publication of a list of possible port calls in an emergency, although Japan is required to do this by international law.

The ability of the ship *Akatsuki Maru* to withstand collision, the design and survivability of the plutonium casks in a collision or to high pressure in the event of sinkings, and their resistance to fire has also been kept secret. Much to the annoyance of most countries along the possible routes, the Japanese government has refused to reveal any relevant safety information. Australia, because she exports uranium to Japan, is the only country to condone these shipments officially, although they pass close to our southern shores.

If the ship had to dock in an emergency, any port along the route would be vulnerable. If 1.7 tons of plutonium burned, depending on wind currents, most inhabitants of the globe could be in peril. A Greenpeace ship tracking the *Akatsuki Maru* collided with its escort boat just out of Cherbourg on November 9, 1992.

Thirty tons of plutonium will be shipped by the year 2010, with possibly 150 tons over the next thirty years—enough for at least 20,000 bombs. So sabotage, piracy, or diversion are real possibilities, particularly in this unstable world. The plutonium ship could also be attacked by a small missile, which is easy to obtain these days on the international arms market.

Ship collisions occur most frequently in crowded waterways, and the plutonium ship may need to traverse the Panama or Suez canals, or sail through the crowded Straits of Malacca. It is a double-hulled ship, constructed not to sink. But the *Titanic*, which was compartmentalized, was also built never to sink.

Because Japan's breeder reactors will not be commercially viable for another twenty to thirty years, an estimated surplus of seventy tons of plutonium will be available by the year 2010, a "stockpile which will surpass in the 2020s all the military plutonium ever produced by nuclear superpowers for weapons use," according to Juzaburo Takagi, a leading Japanese nuclear opponent.

Japan has the technology to produce nuclear weapons within weeks if she so desires. This is called lateral proliferation—aided and abetted by the United States toward a former violent enemy.

Japan will use mixed oxide fuel (MOX), which is mixed plutonium and uranium. It is five to fifteen times more expensive than uranium 235, and is much more reactive and dangerous. As I have stated previously, uranium is now in plentiful supply because of a world glut, and dry storage of spent fuel rods is much cheaper than reprocessing. The only reason, therefore, to stockpile plutonium is for the production of thermonuclear devices or nuclear bombs.

Options for the Japanese nuclear weapons have been discussed publicly by politicians and intellectuals for years.[6]

The plutonium slated to fuel future breeders now lies in the thousands of spent fuel rods stockpiled in storage pools throughout the world. Before plutonium can be used in a breeder, however, it must be separated from the hundreds of other waste fission products found in the spent rods. In "reprocessing," the rods must first be cooled in spent fuel pools for several years, to allow the radioactive materials with shorter half-lives to decay; the "hot" rods would then be packed into strong lead containers, loaded onto trucks, and shipped over miles of highway to a reprocessing plant, where they would be dissolved in vats of nitric acid. The plutonium and uranium would then be separated to be purified, leaving behind the remaining high-level fission waste in solution. The purified plutonium and uranium would be stored in heavy protective containers, to be shipped to fuel fabrication and breeder reactor sites.

Reprocessing expands the volume of radioactivity in one irradiated fuel rod by 200,000 times and results in a liquid, which is a less mechanically stable form of waste than a solid fuel rod.

The U.S. military has used this reprocessing technique to obtain weapons grade plutonium since the 1940s. Until recently, tons of extracted plutonium were routinely shipped along the nation's highways and railways in unmarked trucks. Military reprocessing ceased in the late 1980s, but some plutonium will still be moving along the highways from decommissioned weapons.

The operation of a breeder reactor is much more hazardous than that of an ordinary commercial reactor. The core is made of plutonium 239, surrounded by a blanket of uranium 238 which captures neutrons and converts to more plutonium. So as you burn plutonium, you make more than you lose. Once out of control, a fission reaction in a breeder could cause not only a meltdown but also a fully fledged nuclear explosion. (The nuclear industry calls this a rapid disassembly accident.)

In addition, the breeders are cooled with liquid sodium (rather than water), a substance that ignites spontaneously when exposed to air and is therefore highly dangerous in its own right. In 1966, a near meltdown at the Enrico Fermi I breeder in Michigan threatened the lives of thousands of Detroit-area citizens; following the accident, the reactor was shut down.

The American moratorium on the operation of commercial reprocessing plants and breeder reactors, declared by President Carter in 1977, failed to recognize that such nuclear facilities are mere refinements: since anyone with appropriate knowledge, working in a small properly outfitted laboratory can extract enough plutonium from spent fuel rods to build an atomic bomb, every commercial reactor breeds plutonium that could be used for nuclear weapons.

In the 1970s, in a joint announcement made by the American Electric Power Research Institute (a research arm of the electric utility industry) and Britain's Atomic Energy Authority, scientists from both countries admitted for the first time that current nuclear reactors are producing vast quantities of waste that could be fabricated into bombs. More specifically, they acknowledged that as the radioactivity of most isotopes in spent fuel rods lessens over time, every spent fuel storage facility in the world becomes an increasingly accessible "plutonium mine." To deter non-nuclear countries and terrorists from taking advantage of this situation, these scientists advocated a new method of reprocessing. Called the "Civex" method, it is supposedly more effective than the previous "purex" technique because, instead of individually separating and purifying the uranium and plutonium, it combines the two elements with other fission products to create an impure mixture that is too radioactive to handle safely. Civex advocates were convinced that this new technique would discourage proliferation, but it is doubtful. To date, Civex has only been conducted on benchtop apparatus in a laboratory.[7] And obviously this Civex method makes little difference to the tons of reprocessed plutonium already available for weapons production.

Since the cold war has ended, and the CIS economy falls to shreds, disenfranchised scientists are using ingenious methods to make money. In November 1992, the British newspaper, *Sunday Express* said it had uncovered and foiled a plan to sell Iraq large quantities of weapons-grade plutonium worth $80 million. If the deal had transpired, Hussein might have been able to build twenty atomic bombs within two years.[8]

The ability to create plutonium by means of nuclear fission is one of humanity's most diabolical powers. Once created, this isotope must be isolated from the environment virtually forever. However, some plutonium—perhaps as much as two percent or more—cannot be accounted for; it presumably escaped into the ecosphere during reprocessing, transport, and other industry activities. Or was it stolen? It has been estimated that by the year 2000, 1,139 tons of plutonium will have been produced from global nuclear power plants alone, rising to about 2,100 in the year 2010. About 62 to 75 tons will be discharged annually from nuclear power plants between now and the year 2000. Add to these figures 257 tons of plutonium in the nuclear weapons of the world—CIS, 125 tons; United States, 112 tons; UK, 10.5; France, 6; China, 2.5; Israel, 0.25; and India, 0.3 tons.[9] Remember that one pound is sufficient, hypothetically and if adequately distributed, to kill every person on earth. Of course, plutonium is not evenly distributed like this.

At least fifty pounds of plutonium are buried or scattered in fine particulate form over a large area called Maralinga in Central Australia, where the British were invited by our government to test their weapons. This is the tribal homeland of a small group of the Tjarutja aborigines, who were removed from the test site but now inhabit the area again.[10] In an article in the *New Scientist* in June 1993, Britain may have underestimated this contamination by a factor of ten. Thirty tests using conventional explosives in conjunction with plutonium caused the worst contamination, and twelve of these sent jets of molten plutonium up to 1,000 meters into the air.[11]

Plutonium pollution is not just a problem of the future. It affects us all right now. In 1975, a study carried out by the National Center for Atmospheric Research in Boulder, Colorado, revealed that more than five metric tons of plutonium were thinly dispersed over the earth as a result of nuclear bomb testing, satellite reentries and burnups, effluents from nuclear reprocessing plants, accidental fires, explosions, spills, and leakages. The Chernobyl accident added half a ton to this inventory.[12] And the planned payload of the next Challenger after the Shuttle accident of 1986 was to have been 47.6 pounds of plutonium 238.[13]

Orbiting nuclear powered spacecraft now number forty-two, some powered by plutonium, and some uranium. Most travel in orbits populated with large quantities of space debris. A collision could send radioactive debris into the atmosphere that would eventually rain down on earth.[14]

As a result of these activities, most males in the Northern Hemisphere already carry a very small plutonium load in their reproductive organs. As plutonium contamination of populated areas worsens, that load will increase, with potentially devastating consequences for their offspring.

Chapter 6

M.A.D.: Mutually
Assured Destruction

In 1965, the U.S. government predicted that one thousand nuclear reactors would be in operation by the year 2000. However, because of the massive expense of construction and operation combined with public opposition, no new reactors have been ordered since 1974. Even so, at this time, 110 nuclear reactors are on line in the United States, and the nuclear industry and the Department of Energy would like to construct about 175 advanced "passively safe" reactors over the next thirty years.[1] Somehow they have managed to quiet public concern and their own fears and trepidation after Three Mile Island and the tragedy of Chernobyl, and convince themselves that nuclear power is a viable proposition. This notion has yet to be tested before the American public, but the industry is presently spending millions of dollars in propaganda and expensive ads to teach the people that nuclear power is the answer to greenhouse warming.

Because of the dramatic decline in the U.S. market for reactors over the last several decades has threatened the financial survival of the nuclear industry (no nuclear reactors have been built since 1972), multinational nuclear suppliers, especially Westinghouse and General Electric, have heavily promoted their overseas trade. Some of their most eager customers have been developing nations, such as the Philippines, South Korea, Mexico, Spain, Taiwan, and Yugoslavia. Many of these countries lack the capital to purchase a nuclear power plant, so American loans were and are arranged through the World Bank.

While these countries sometimes lack power-transmitting grids to distribute the electricity generated, their desire to produce electricity is not always their primary motivation. Often their ultimate goal is to gain access to nuclear weapons grade materials and to join the "nuclear club."

India proved this point in 1974 when she tested an atomic bomb; Israel owns hundreds of neutron bombs and tactical and strategic nuclear weapons.[2] Israel cooperated with South Africa to test its first nuclear weapon in 1979; the Shah of Iran who commanded one of the world's largest air forces, planned to have 20 reactors in operation by 1994, and now Iran is on the verge of building nuclear weapons aided and abetted by Argentina and China together with help from the former Soviet Republics.[3] Argentina, Brazil, South Korea, Taiwan, Iraq, Cuba, the Philippines, Mexico, and Japan all have the potential to develop nuclear weapons in the very near future.

Pakistan now admits that it has nuclear weapons capabilities, and North Korea has the reactors and facilities to develop weapons, much to the extreme discomfort of Japan who is also capable of constructing nuclear weapons at short notice.[4]

In December 1991, President Bush announced that all U.S. nuclear weapons had been withdrawn from South Korea, which could itself now manufacture nuclear weapons because reactors were sold to it by the United States. So one door of hope opens while another closes.

When I visited Prague in August 1991, to address an Eastern Bloc conference on nuclear power, I was amazed to find that some of the Czechoslovakian bureaucrats had not been informed about Chernobyl by the former Soviet Union, and that they were completely ignorant about its ramifications. They had been extensively lobbied by German, French, and U.S. nuclear corporations to build reactors in their newly liberated country. They were resistant to my talk but they clearly had never heard of the medical implications of nuclear technology before.

In Hungary, ever since its liberation, people such as Edward Teller, a native son representing the U.S. nuclear industry, have made numerous visits to convince their colleagues that nuclear power is the answer to Hungary's future energy needs. The corporations making such representations are Siemens KWU of Germany, Electricité de France, Westinghouse, General Electric, General Atomics, and Bechtel of the United States, and Atomic Energy of Canada Ltd.

These firms have showered Hungary and Czechoslovakia with feasibility studies, offers, and bids to build nuclear facilities of every type, and have promised tens of thousands of jobs, and billions of dollars of investment to people who do not understand the concept of money or capitalism.

While the present old Soviet-built reactors are experiencing accidents and radioactive leaks, the construction of more hugely expensive units could well break the backs of these new revolutions. The lack of energy efficiency in these countries is appalling, and energy conservation and investment in small scale hydroelectricity, as practiced before World War II, would provide a large part of the solution. Tragically, a large new reactor in Slovakia will provide power for a giant aluminum plant—one of the country's worst polluters.[5]

This spread of nuclear power plants around the world—and the directly related proliferation of nuclear weapons and waste—still seriously threatens global peace and order. At this writing, the United States and the CIS still maintain the balance of power, but the sale of each new reactor tips the scale toward a world of uncontrollable proliferation, in which regional nuclear conflicts could draw the superpowers into all-out nuclear war.

The nuclear industry knows that the reactors it sells produce material for weapons, but its major concern seems to be corporate profit, not morality or human survival. (General Electric is known to have conducted promotional conferences with Egypt and Israel on the same day.)

Nuclear suppliers have, however, voiced concern over the use of reactor by-products for military ends: corporate representatives from Great Britain, the United States, France, West Germany, Japan, Sweden, and other countries agreed in 1978 that any country buying a nuclear reactor and using its plutonium to manufacture bombs would receive a "reprimand." Such a scolding would not, of course, preclude further sales to the country at fault.

The United Nations is keenly aware of the weapons potential involved in nuclear power. To minimize the diversion of nuclear materials toward weapons manufacture, it passed the 1968 Non-Proliferation Treaty (NPT), whose signators agreed not to utilize nuclear materials to build bombs, or to sell such materials to any other country for that purpose. But the NPT is impotent: the world's nations are not required to ratify it, and those that have can retract their ratification with ninety days' notice—during which time a nuclear weapon can be produced.

In 1965, the United Nations established the International Atomic Energy Agency (IAEA) to police the world's nuclear facilities and deter the conversion of fissionable material into weapons. Since 1968, the purpose of the IAEA has been to enforce the NPT, but it has been given few powers with which to carry out its responsibilities. Agency inspectors (all too few in number) are authorized only to inspect for the misappro-priation of nuclear materials for use in bombs; they are not empowered to prevent a country from actually building a bomb. Moreover, this purportedly neutral regulatory body has promoted reactor sales: in 1975, it recommended that Pakistan build twenty-four nuclear plants by the end of the century.

To illustrate the inadequacy of these safeguards, North Korea finally signed the NPT in 1985, but it only allowed IAEA inspectors into the country to ensure compliance in 1992. The planned construction of new reactors and the com-pletion of a reprocessing plant at Yongbyon could produce over 200 kilograms of plutonium per year by the late 1990s.[6] In October 1992, the then U.S. Secretary of Defense, Dick Cheney, said, "We have good reason to believe that the North Koreans are aggressively seeking to develop nuclear wea-pons."[7]

In fact, in March 1993 North Korea announced its intention to withdraw from the 1968 Non-Proliferation Treaty, which would have come into effect on June 12, 1993. Subsequently, due to intense pressure from the United States, North Korea temporarily suspended its withdrawal from the treaty.[8] It has also violated its obligations to allow IAEA inspectors to visit suspected nuclear weapons sites in April 1993.[9]

If North Korea developed nuclear weapons, South Korea, Taiwan, and Japan would also be surely tempted to build their own. Canada agreed to sell uranium to Taiwan, which has not signed the NPT. The United States, which has a bilateral agreement with Taiwan, and the World Bank on May 31, 1993, approved a $165 million loan to upgrade Iran's electric power system—presumably nuclear, in an oil-rich country. U.S. Intel-ligence estimates Iran could build a nuclear weapon by the year 2000.[10]

The American government is also conscious of the "diver-sion" problem. The Nuclear Non-Proliferation Act, passed by

the African National Congress and Nelson Mandela, the Tiger
guerrillas kill hundreds in Sri Lanka, and Congress still chan-
nels huge quantities of money and weapons through the CIA
to right-wing soldiers in Haiti, Guatemala, El Salvador, Nicara-
gua, Peru, Colombia, and the rest. This is just to name a few
of the killing fields on the planet today.

But people can't kill without guns, bombs, missiles, planes,
and grenades. Where do the arms come from? Obviously from
the CIA and from indigenous armies. But where else?

In 1974, after the Middle East oil embargo, weapons became the major source of international currency, replacing oil
and gold. Since then the world has gone mad. The major
weapons suppliers are France, China, the CIS, and the United
States, with healthy support from Israel, Germany, South
Africa, Brazil, and Australia, among others. The weapons trade
is so lucrative that countries can't afford not to be involved.

In 1992, little Australia with a population of 18 million, was
earning $1.5 billion a year from the sale of heavy military
equipment, a figure that has tripled since 1982–1983, proudly
reported by the Minister for Defense, Science, and Personnel,
Mr. Gordon Bilney.[15]

But in 1991, for the first time since 1983, the United States
surpassed the former Soviet Union as the biggest supplier of
weapons to the third world, reaching a record $18.5 billion, up
from $7.8 billion in 1989. The United States sold $30.7 billion
worth of weapons to the Middle East between the years 1987
to 1990, while Soviet sales were $17.5 billion for the same
period. Ironically, Saudi Arabia and Iraq were, by a wide mar-
gin, the top purchasers between 1983 and 1990.[16] The question
about the source of Iraq's military might is therefore redundant.

And in 1991, the United States agreed to deliver over $63
billion in arms sales to the Middle East. Furthermore, the
United States has deliberately castrated the "Big Five" talks
curbing international arms sales in the wake of the Gulf War.

One U.S. negotiator attached to the talks described the material produced as "completely nonbinding stuff." He said: "We agreed to nothing just for the sake of having something to agree to."

Meanwhile, Washington and Paris announced sales of F-15s, F-16s, and Mirages to Saudi Arabia, and George Bush, to boost his flagging campaign, announced the sale of 150 F-16s to Taiwan—with the prearranged support of Bill Clinton. China was very annoyed. A most likely outcome of continued U.S./third-world arms sales is to remove incentive for constraint by any other weapons vendors.[17]

The General Accounting Office concluded in 1991 that the United States is not adequately monitoring the arms sales that it makes, thereby losing track of billions of dollars in arms to other countries. Weapons were sold in contravention to U.S. law or had simply disappeared from the custody of foreign powers. Philippino officers had falsified requisitions in order to steal and sell M1 rifles, four hundred rounds of ammunition and two hundred grenades probably to anti-government insurgents. This example must surely be commonplace throughout the third world. In 1988, developing countries spent $216 billion on their armed forces, almost 5 percent of their gross national product. And that was a relatively light year for third-world arms spending.[18]

Killing and power comes before food and care. In the first world, corporations make money by selling death with absolutely no moral scruples. So millions of innocent people, mostly women and children, die. It is time to put an end to the global killing fields.

No matter where military spending occurs, the effects are the same. Arms expenditures not only enhance the threat of global war but divert precious resources from urgent social needs, obstruct economic growth, fuel inflation, and raise unemployment.

In today's world, 1.5 billion people lack access to professional health services. Over 1.7 billion people have no safe drinking water. More than 730 million people suffer from malnutrition. Although 900 million of their adult citizens are illiterate and more than half of their children do not even attend school, world governments spend twice as much on armaments as on health care.[19]

In the words of Ruth Leger Sivard, former chief of the economics division of the U.S. Arms Control and Disarmament Agency:

There is in this balance of global priorities an alarming air of unreality. It suggests two worlds operating independently of each other. The military world, which seems to dominate the power structure, has first call on money and other resources, creates and gets the most advanced technology, and is seemingly out of touch with those threats to the social order that have nothing to do with weapons. The other world, the reality around us, has a vast and growing number of people living in poverty, old people who need care, more and more children who are unable to attend school, more families who never see a doctor, have an adequate dwelling, or escape from the edge of starvation. The everyday world is a global community whose members are increasingly dependent on one another for scarce resources, clean air and water, mutual survival. Its basic problems are too real, too complex for military solutions.[20]

While the gulf between the rich and the poor continues to widen, the poorest nations of the world continue to arm themselves not only with weapons but also with the capacity to produce nuclear weapons. In 1992, thirty-three countries had nuclear reactors, including fifteen non-nuclear weapons countries.[21]

In 1946, Albert Einstein, apprehensive about the misuse of

the power of the atom, expressed great concern for the future of humanity. Today's nuclear arsenal must exceed his worst nightmare.

Throughout the duration of the cold war, the former Soviet Union constructed a total of 55,000 nuclear weapons and the United States 70,000.[22] But in fact the United States and former Soviet Union together have enough weapons-grade fissile material to make nearly 100,000 nuclear warheads. Uranium obtained from returned warheads is recycled into new nuclear weapons.[23] To use a hackneyed word, this is an unbelievable number, when 100 bombs dropping on 1,000 cities could possibly induce nuclear winter and the end of most life on earth.[24]

We were all lucky to escape annihilation during those tedious hostile years. But although the world feels a safer place, nuclear war could still occur by the accidental launch of a bomb or missile by one of the five major nuclear weapons states, or by a rogue state. Or such a catastrophe could be induced deliberately, either by a deranged military individual, or by a country such as Pakistan or India in a local dispute. In the United States until recently, every year up to 5,000 men who are responsible for nuclear weapons have been discharged from the military because they were either on hard drugs or alcohol, or were mentally unstable. A similar number of deranged individuals would still be responsible in the same way for the CIS weapons. As Admiral Gene La Rocque once said to me, "There is no system fail-safe enough that an intelligent man can't bypass."

A limited nuclear war between small states could still involve the vast arsenals of the five major nuclear nations by accident or design. The technology to initiate a nuclear war is still in place, so even though the perception of the risk is lower, the danger is still very much with us. In fact, United States missiles are still aimed at Russian targets today, even after the end of the cold war and the disintegration of the Soviet Union. According to the *Sydney Morning Herald* (January 14, 1994) an agreement to retarget certain missiles into the open sea has been reached by the U.S. and Russia, but not yet implemented; the military targets in both countries will remain. However, the missiles can be retargeted back within fifteen minutes.

ARMS REDUCTIONS

In 1991, the United States owned 11,400 strategic nuclear weapons (deliverable by rocket or plane). It plans to reduce this number to 3,500 by the year 2003, under the new arms control treaties. It also possessed 8,000 tactical weapons for "short-range" killing, which will be cut to 4,250. The CIS owns 10,600, which are to be cut to 3,000, and 17,000 tactical weapons, to be cut to 4,730.

The reductions in tactical nuclear weapons are based on estimates only by CIS and U.S. officials—there are no specific figures attached to these particular initiatives.

The agreements covering the strategic weapons are the START I and START II Treaties, not yet ratified by the U.S. Senate. START II is still contingent upon Senate ratification of START I. In addition, under START there is absolutely no provision to dismantle any of the bombs or the missiles that transport them. These treaties only require launch tube and silo destruction. Bombers can be maintained but can be converted to conventional weapons. There are absolutely no verification means built into the treaties.

Warheads are to be saved from dismantlement because the United States didn't want the treaty to be verified by allowing the Russians to look at American warheads and thereby assess the total number. The missiles are to be spared because the United States plans to use them for Star Wars practice, while the Russians want to use their missiles for launching satellites into space.[25]

This initiative was instigated by the United States. So, in fact, both the United States and the CIS may actually maintain all the warheads they presently possess, and store them and use them if they so desire. There were two independent initiatives by President Bush in September 1991 and President Gorbachev in October 1991 that planned to destroy some of their tactical weapons. In February 1992 President Yeltsin committed Russia to follow these initiatives promised by Gorbachev. Tactical weapons have never been included in arms control treaties for some strange reason.

So, in truth, the START treaties are vacuous and cruel deceptions upon a world that really believes that true nuclear disarmament is taking place. You can see that the combined arsenals are still very frightening; even with large bilateral reductions within ten years, we still face potential catastrophe.

The internal disruption within the CIS has many nuclear experts very worried. Apparently all tactical weapons have been returned to Russia, but Belarus still houses 100 strategic bombs, Kazakhstan 1,500, the Ukraine 1,700, and Russia 7,300. Technically, the CIS still "owns" and controls these weapons. Belarus and Kazakhstan have ratified START I and the nonproliferation treaty, but the Ukraine, which has possession of two thousand nuclear warheads, is stalling on returning these to Russia and is also claiming possession of 176 strategic missiles and forty nuclear bombs on its territory.[26]

We must pray that state doesn't turn on state and consider using these weapons as a viable option—or sell them.

Although the U.S. ground- and sea-based tactical weapons have been sent home, America still maintains 700 tactical, air-based bombs in Europe. Informed scientists estimate that France owns 600 nuclear bombs, the United Kingdom 300, China 500, and Israel well over 300.[27]

To add fuel to the nuclear fire, President Clinton, at the

instigation of the Pentagon, the DOE, and the State Department, announced in May 1993 that he would renew underground nuclear tests—ending a nine-month U.S. moratorium that followed the lead of Russia and France—on the excuse that he would end all testing in 1996. This move has made the hearts of many good people drop with frustration and anger and could mean that Russia and maybe the Ukraine will resume testing. Due to pressure from Democrats in Congress, it appears very likely that the President will back away from the proposed plan and place a moratorium on further nuclear tests.[28]

The nonproliferation treaty (NPT) comes up for renewal in 1995, and many countries like Iraq and Libya will be so incensed by this action that they, like North Korea, will not renew their commitment to the NPT, thereby making the world a much more dangerous place.[29]

It must be remembered that the United States tested 156 nuclear bombs in the atmosphere of Nevada from 1951 to 1962. The explosions took place only if the wind blew west away from Los Angeles or Las Vegas toward the devout Mormon communities in Nevada, Utah, and Arizona. Each of the pink clouds contained radiation levels comparable to those released over the Ukraine and Europe after Chernobyl. Animals died of acute radiation sickness or were born deformed, and so were patriotic, unquestioning American citizens. Radiation fell across the United States to the East Coast. Hundreds of people have since died of leukemia and cancer in clean-living communities once almost exempt from these diseases. The people were lied to by their government, and the truth has only recently started to emerge over the last few years.[30]

Conventional weapons release the molecular energies of the chemical compound trinitrotoluene (TNT). Nuclear weapons contain the explosive force of the stars: their power can be millions of times greater than conventional bombs.

Only ten to twenty pounds of uranium 235 or plutonium 239 are needed to fuel an atomic bomb. Like a nuclear reactor, such a bomb operates on the fission principle: enough of the material is brought together to form a critical mass, resulting in a chain reaction; in one millionth of a second all the nuclei in the mass decompose, liberating the explosive power of as much as 20,000 tons of TNT. The explosions of one of these bombs over a modern city can kill 100,000 people and lay waste to an area miles in diameter.

A hydrogen bomb works on the fusion principle, the same process that fuels the sun. In such a bomb the atoms in one thousand pounds of lithium deuteride are joining two at a time to form atoms of helium. In the process, enormous concussive forces and thermal energy are released. The high temperature needed to get the process going is provided by an atom bomb, which serves as a triggering mechanism, releasing more than 15 million degrees of heat. A thousand times more powerful than an atom bomb, one hydrogen bomb can kill millions of people within seconds.

By adding a 1,000-pound shell of uranium 238 to the deuterim and uranium 235 (or plutonium 239) in the average hydrogen bomb, the weapon's explosive power may be doubled or tripled at very little cost. The result is a fission/fusion/fission superbomb with the explosive power of over 20 million tons of TNT. In addition to the great force and heat it gives off, such a bomb creates an enormous number of fission products, which cause radiation effects long after detonation.

The explosive force in one million tons of TNT is called a megaton. During all of World War II, a cumulative total of three megatons were detonated. Today, some hydrogen bombs have the explosive power of five to twenty-five megatons. The detonation of a single weapon of this nature over any of the world's major cities would constitute a disaster unprecedented in human history.

Today, many missiles carrying one heavy warhead have been replaced by more accurate ones carrying three to ten lighter warheads; called MIRVS (multiple independently targetable reentry vehicles), these weapons are capable of breaking away from the main rocket and landing on separate targets with deadly accuracy.

Tens of thousands of these nuclear bombs can be released in a matter of seconds—possibly by accident, as John F. Kennedy once warned. There have been hundreds of near accidents over the past forty-six years. On a number of occasions, U.S. nuclear armed submarines on reconnaissance have collided with Soviet vessels; a Russian aeroplane carrying a nuclear weapon crashed into the Sea of Japan; a Soviet guided missile destroyer reportedly exploded and sank in the Black Sea; and an American aircraft is known to have accidentally dropped four plutonium bombs on Spain that did not detonate.*

Our future is vulnerable not only to human error, but also to the frailty of human emotions. Although the CIS and the United States are now friends, nobody knows what the future holds. A reckless national leader, or one under severe stress, could be disastrous. Several weeks before President Nixon resigned, concerned administration officials removed the mechanisms by which he could start a nuclear war. Premier Leonid Brezhnev was treated with cortisone, a drug that occasionally induces acute psychosis. And I spent one and a quarter hours with President Reagan in the White House, in an intense one-to-one conversation on nuclear weapons, nuclear strategies and technologies, and the probability of nuclear war. He

---

*The United States dug up tons of the heavily contaminated soil and buried it in trenches in Barnwell, South Carolina, where the average monthly rainfall of over four inches has probably leached some of the buried plutonium into the Savannah River.

was very apprehensive throughout because his knowledge on these subjects was almost zero, and I intermittently found myself holding his hand to reassure him through his moments of tension.

The complex technology required by America's military and defense apparatus concentrates an enormous amount of power in the hands of a few common mortals. Spending long hours in cramped quarters, two men used to guard each Titan missile silo in the United States, knowing that at any moment they could receive orders to launch the missiles against the enemy. Each man was armed with a pistol and was ordered to shoot the other if he exhibited abnormal behavior. Since these missiles were first deployed, according to a report published by the National Academy of Sciences, thirty of these men are known to have been seriously psychologically disturbed.

Now in 1992, the Titans have gone, but instead five hundred Minutemen II missiles (each armed with three bombs) and fifty MX missiles (each with ten bombs) are still on full alert status. In addition, the last three Trident submarines are under construction at a cost of $3 billion; and the D5 missile for the Trident will continue to cost U.S. taxpayers $15.7 billion. There are enough warheads on each Trident submarine to destroy most major cities in the Northern Hemisphere.[31]

The twenty B2 bombers, which are now totally obsolete, have cost $44.4 billion so far, and fifteen have already been constructed. Over the last nine years since Reagan's Star Wars speech, $30 billion has been spent, and Congress continues to allocate billions more dollars each year for this futile notion.[32]

The cold war is over and this money must now go to the people of the United States for health care, education, and environmental protection.

But there are even more dangerous plans afoot. In March 1992, a Defense Planning Guidance for fiscal years 1994 to 1999 was leaked to the New York Times by an official who believed the post-cold war strategy should be debated in public. This plan basically says that the United States should be the world's only superpower, whose leaders "must maintain the mechanisms for deterring potential competitors from even aspiring to a larger regional or global role" i.e., the New World Order. The United States will cooperate with the United Nations only if it is in America's interest, and it will feel free to use nuclear, chemical, or biological weapons to prevent the production of weapons of mass destruction in other countries, and also in conflicts that otherwise do not directly engage U.S. interests.[33] And in January 1992, advisors urged then President Bush to target every reasonable adversary around the globe with nuclear or non-nuclear weapons, recommending that five thousand nuclear weapons be targeted at potential enemies over the next few years,[34] while President Clinton's Secretary of Defense, Les Aspin, announced that the United States should and could preserve its ability to fight and win two regional wars at once.[35] In this context, I must say that I am extremely concerned about the almost unilateral role that the United States is currently playing in the United Nation peacekeeping forces. It seems to be trigger-happy and uses its high-tech weapons at will. My sense is that because the United Nations is very short of cash, the United States takes advantage of this and steps into the financial void with its own policies and weapons, almost using the noble aspirations of the United Nations as a camouflage for its own not-so-hidden agenda.

The election of President Clinton may or may not change these surreptitious and extraordinarily arrogant attitudes—it remains to be seen; but scrupulous vigilance on our part is advisable.

There very nearly was a nuclear war on November 9, 1979. A fellow in the Pentagon plugged a war games tape into a supposedly fail-safe computer and the computer took it for real. All the American early warning systems around the world went on alert for six minutes. Three squadrons of planes took off armed with nuclear weapons. At the seventh minute the presidential 747 command post was readied for takeoff. (They couldn't find the president. He was to be notified at the seventh minute). If in twenty minutes it hadn't been stopped, we wouldn't be here right now. And twice again, on June 3 and June 6, 1980, computer errors nearly led us into a nuclear war. Remember, twenty minutes is currently the time limit for a retaliatory nuclear attack.

In 1983, when the U.S./NATO forces were conducting a very aggressive war game against the Soviet Union, against a psychological background of President Reagan's "Evil Empire" and other belligerent statements, the Soviet leaders thought they were about to be attacked. A double agent in London contacted Prime Minister Thatcher, who called President Reagan in the middle of the night, to warn him that if the exercise was not cancelled, the Soviets would initiate a nuclear attack within hours. This he did, his rhetoric subsequently changed, and we are still here to tell the tale.[36]

Now the new era of peacemaking between the United States and the Confederation of Independent States makes these events unlikely to be repeated, but other international dynamics may take their place unless we eradicate nuclear weapons from the face of the earth.

THE MEDICAL CONSEQUENCES OF NUCLEAR WAR

What would happen if the world's nuclear arsenal were put to use?

Erupting with great suddenness, a nuclear war would probably be over within hours. Several hundred to several thousand nuclear bombs would explode over civilian and military targets in the United States (every American city with a population of 25,000 or more may well still be targeted despite the end of the cold war), and an equal or greater number of bombs would strike the principal targets in Europe, the Confederation of Independent States, and China (the latter two countries are still targeted by the United States). Both major and minor population centers would be smashed flat. Each weapon's powerful shockwave would be accompanied by a searing fire ball with a surface temperature greater than the sun's that would set fire storms raging over millions of acres. (Every 20-megaton bomb can set a fire storm raging over three thousand acres. A 1,000-megaton device exploded in outer space could devastate an area the size of six western U.S. states.) The fires would sear the earth, consuming most plant and wild life. Some experts believe that the heat released might melt the polar ice caps, flooding much of the planet. Destruction of the earth's atmospheric ozone layer by the rapid production of nitrous oxide would result in increased exposure to cosmic and ultraviolet radiation.

People caught in shelters near the center of the blast would die immediately of concussive effects or asphyxiation brought on as a result of oxygen depletion during the firestorms. Exposed to immense amounts of high-energy gamma radiation, anyone who survived near the epicenter would likely die within two weeks, of acute radiation illness.

Those who survived, in shelters or in remote rural areas, would reenter a totally devastated world, lacking the life-support systems on which the human species depends. Food, air, and water would be poisonously radioactive. Physical suffering would be compounded by psychological stress. For many, the loss of family, friends, and the accustomed environment would bring on severe shock and mental breakdown. It should also be noted that many plants and animals are more sensitive to radiation than human beings. They would also die.

In the aftermath, bacteria, viruses, and disease-bearing insects—which tend to be thousands of times more radioresistant than human beings—would mutate, adapt, and multiply in extremely virulent forms. Human beings, their immune systems severely depleted by exposure to excessive radiation, would be rendered susceptible to the infectious diseases that such organisms cause: plagues of typhoid, dysentery, polio, and other disorders would wipe out large numbers of people. Millions of dead human and animal bodies would be potent breeding grounds for bacteria, which would also propagate disease amongst the unlucky survivors.

The long term fall out effects in the countries bombed would give rise to other epidemics: within five years, leukemia would be rampant. Within fifteen to sixty years, solid cancers of the lung, breast, bowel, stomach, and thyroid would strike down survivors.

Exposure of the reproductive organs to the immense quantities of radiation released in the explosions would result in reproductive sterility in many. An increased incidence of spontaneous abortions and deformed offspring, and a massive increase of both dominant and recessive mutations, would also result. Rendered intensely radioactive, the planet Earth would eventually become inhabited by bands of roving humanoids— mutants barely recognizable as members of our species.

But other scenarios conjure up a freezing nuclear winter, when clouds of black, toxic radioactive smoke would block out the sun for a period of one year or more—and most living organisms that happened to survive the initial blast and radiation would freeze to death in the dark. It is not clear, but it is possible that all life on earth would be exterminated.

What would be left? Experts have projected two possible

scenarios. According to one, hundreds of millions of people in the targeted countries would die, but some might survive. According to the second, the synergistic ecological effects of thermal and nuclear radiation, long-term fall out, exposure to increased cosmic radiation and to nuclear winter make destruction most likely absolute. There would be no sanctuary.

Is it not remarkable how we manage to live our lives in apparent normality, while, at every moment human civilization and the existence of all forms of life on earth are still threatened with sudden annihilation? We seem to accept this situation calmly, as if it were to be expected. Clearly, nuclear warfare presents us with a specter of a disaster so terrible that many of us would simply prefer not to think about it. But soothing our anxieties by ignoring the constant danger of extinction will not lessen that danger. On the contrary, such an approach improves the chances that eventually our worst fears will be realized.

Nuclear disarmament is still the first and foremost task of our time; it must be given absolute priority and we must not be placated by "large reductions" in missile silos and launch tubes. All the actual warheads and bombs must be dismantled within the next several years. Moreover, with nuclear reactors and thousands of containers of radioactive wastes vulnerable to attack around the world, all war—conventional or nuclear is rendered obsolete.

Our environmental circumstances changed dramatically when the appearance of nuclear weapons forever altered the nature of war. If we are to survive, we must accept personal responsibility for war and peace. We cannot afford to delegate these responsibilities to generals, polititians, and bureaucrats who persist in the politics of confrontation and in outmoded ways of thinking that have always caused—and never prevented—war. International disputes must now be settled by reason, not with weapons. The current war in Bosnia illustrates the bestial madness of war. They haven't yet attacked a nuclear reactor, but such a situation is possible.

Only if we abolish nuclear weapons and permanently halt the nuclear power industry and contain radioactive waste inventories from entering the biosphere, can we hope to survive.

To achieve these ends, it is vital that people be presented with the facts. Today more than ever, we need what Einstein referred to as a "chain reaction of awareness": "To the village square," he wrote in 1946, "we must carry the facts of atomic energy." Once presented, the facts will speak for themselves.

Out of the growing number of organizations opposed to nuclear power and nuclear arms must come a grassroots movement of unprecedented power and determination. Its momentum alone, will determine whether we and our children, and all future generations of humankind, and perhaps even life itself, will survive.

Chapter 7

Three Mile Island

A t four A.M. on March 28, 1979, the worst accident in the history of the U.S. nuclear industry began and, within several hours, the badly damaged radioactive core of the nuclear reactor at Three Mile Island in Harrisburg, Pennsylvania, had partially melted down. Although later engineering analysis doubted that a true "China Syndrome" could have taken place because of the hard rock formation beneath the reactor, it is certainly well within the realm of possibility that a substantial fraction of the reactor's radioactive inventory was released into the atmosphere.

A normally operating 1000-megawatt reactor contains fifteen billion curies of radiation, the equivalent of the long-lived isotopes released by the explosion of 1,000 Hiroshima bombs. When an atomic bomb explodes, much of the fall out is forced into the stratosphere and circulates west to east around the globe. Fall-out is intermittent, occurring only with low-pressure atmospheric conditions. However, in the case of a meltdown, any damage incurred to human population and property would depend upon the prevailing wind direction at the time of the release of the radioactive plume, and the plume would hover close to the earth depositing fall out continuously along its path.

To understand the awful magnitude of destruction from a meltdown, the government-sponsored Rasmussen Report, as updated by the Union of Concerned Scientists, is very helpful. It states that in the "worst possible case" (of an assumed ten million people at risk for a single accident), 3,300 people would die from severe radiation damage within several days; 10,000 to 100,000 people would develop acute radiation sickness with two to six weeks of initial exposure; 45,000 would become short of breath since the intensely radioactive gases produce lung damage; 240,000 others would develop acute hypothyroidism with symptoms of weight gain, lassitude, susceptibility to cold, impaired and slow mental functions, loss of appetite, constipation, and absent menstruation; 350,000 males would become temporarily sterile as gamma radiation damages sperm; 40,000 to 100,000 women would cease to menstruate, many permanently. In the fetal population, up to 100,000 babies would be born cretins with mental retardation because radioactive iodine would destroy the thyroid gland; 1,500 others would develop microcephaly (small heads) since the developing central nervous system is highly susceptible to the deleterious effects of radiation. There would be 3,000 deaths in utero. Five to sixty years later, cancer would develop in various body organs of 270,000 people and there would be 28,800 cases of thyroid malignancy.

Economically, a meltdown would cost tens of billions of dollars in property damage, but even more horrifying is the fact that it could contaminate an area the size of Pennsylvania for hundreds to thousands of years making it uninhabitable.

Over thirty million Americans presently live within thirty

miles of a nuclear reactor. How many could be evacuated?

At the time of Three Mile Island (TMI) the existence of an effective local government (state and county) evacuation plan was not a prerequisite for granting an operating license for a reactor. Under current regulations, individual states submit emergency plans to the Nuclear Regulatory Commission (NRC) for approval, but at the time of the accident only a few states had NRC-approved plans, and Pennsylvania was not among them.

The size and location of the population to be evacuated would depend upon the prevailing wind conditions, but the evacuation of a large population—New York City and surrounding areas, for instance, within hours after a meltdown— would be a frightening scenario.

Even if such an evacuation were possible, physicians are totally unprepared to cope with such a large population of contaminated patients. Presently, our medical education does not include the decontamination and treatment of radioactive patients; neither is the issue addressed in medical journals. Government regulations state that a patient should be hosed down to remove the isotopes from her skin and the contaminated water should then be drained. But it would be very dangerous for medical personel to handle contaminated patients because they themselves would become contaminated. The sorrowful truth is that we cannot decontaminate radioactive lungs or other body organs, nor can we treat organs acutely damaged by radioactive exposure. We cannot cure sterility, and those males who might regain their fertility would have to be strongly advised not to reproduce since their sperm would almost certainly have sustained some genetic damage from radiation exposure. The medical community has made some progress in recent years in cancer treatment, but most forms of adult cancer are still incurable.

We know of only a few varyingly effective modes of medical

action to follow at the time of a meltdown. First of all, it is advisable for all people living within a hundred mile radius of a reactor to have a supply of inert potassium iodide tablets in their medicine cabinets, available for use in such an emergency. If a medically prescribed dose were taken thirty minutes to several hours before exposure, the nonradioactive iodine would be absorbed by the thyroid gland and subsequent uptake of radioactive iodine by the thyroid would be blocked. For exposed, unprotected patients who develop acute hypothyroidism, a thyroid extract could restore normal endocrine balance. A much less effective treatment for those who had no medication available could be to take a laxative immediately. By reducing transit time through the gut, absorption of the radioactive isotopes could be impeded. In short, we are hardly prepared to face such an emergency.

How Did Three Mile Island Happen at All?

The accident began with a mechanical failure and the automatic shutdown of the main feedwater pumps in the secondary coolant system, which closed some valves, thereby causing the water in the primary coolant system covering the radioactive core to overheat. The pressurizer operator relief valve (PORV) in the primary system automatically opened as it was designed to do and started draining precious cooling water away from the core. When the accident began, the reactor scrammed and control rods automatically dropped between the uranium fuel rods in the core, stopping the fission reaction. The core, however, remained exceedingly hot, and the light on the operating panel failed to indicate that the valve was not shut but stuck open. This valve remained open for approximately two hours before the operators understood the situation. Thinking the core was covered with water, the operators turned off the water flow from the emergency cooling system, causing the water to rapidly drain from the core through the PORV while the emergency cooling system was closed tight.

Then, one hundred tons of uranium core overheated and the zirconium cladding of the rods reacted with water at 2,200 degrees Fahrenheit, melted, and exposed the cooling water to highly radioactive fission products and alpha emitters. The core was severely damaged and the hydrogen gas that had collected in the containment vessel spontaneously exploded several hours after the accident began, exposing the containment vessel to potentially dangerous explosive forces that could have weakened it. Believe it or not, this incredibly serious event was not recognized by either the operators, the utility, or the NRC for several days. As more hydrogen continued to be generated over following days, Chairman Hendrie of the NRC and others feared that oxygen released by radiolysis from water molecules would unite with the hydrogen bubble, explode, and possibly rupture the containment vessel releasing large quantities of radiation; or that a noncondensible hydrogen bubble would eventually uncover the core leading to a meltdown. Meanwhile, in order to reassure the American people that the reactor posed no danger, President and Mrs. Carter were taken into the operating room of TMI Unit 2 on the afternoon of April 1, at the height of the anxiety about this hydrogen bubble. This foolish exercise exposed the president to unnecessary danger.

Throughout the accident, the highly radioactive cooling water was being pumped through the PORV onto the floor of the reactor and from there into a tank in an adjacent auxiliary building. At this time, large quantities of radioactive gases were vented from this auxiliary tank from a leaking valve into the atmosphere, so intermittent fumes of radioactivity were dispersed from the reactor. The wind blew mainly to the northwest quadrant. The radiation was measured by various gamma-counting devices. Beta and alpha radiation were not measured, but it has been estimated that 2.4 to 13 million curies of noble gases—Xenon and krypton—were released, and 13 to 17 curies of radioactive iodine escaped. Although noble gases do not combine chemically in the body, they are absorbed by the lungs after inhalation. Ten times more fat-soluble than water-soluble, they tend to concentrate in the abdominal fat pad and the fat of the upper thighs. Xenon 133 and krypton 85 are high-energy gamma emitters like X rays. Thus the reproductive organs of people living near TMI could well have been exposed to gamma radiation, particularly if those people were immersed in clouds of noble gases.

Because the monitoring device in the auxiliary building stack went off scale early in the accident and was off scale for forty-eight hours (it was not designed to measure such large quantities of radiation), estimates of radiation release were extrapolated from radiation monitors some distance from the stack. Therefore no hard data exists concerning the absolute quantities of the radiation release. Furthermore, because alpha and beta radiation were not routinely measured, it is impossible to precisely know which radioactive elements escaped. Noble gas measurements were not begun until April 5th—eight days after the accident began. In other words, all calculations are only educated guesses from the available but inadequate data. On day three of the accident, 172,000 cubic feet of radioactive water was discharged by the utility into the Susquehanna River without NRC permission, to make room for more highly radioactive water from the containment building. Official estimates from the government-sponsored Kemmeny Commission also predict that less than one case of cancer will occur from the radioactive releases in the exposed population during their lifetimes. However, using the Hanford/Mancuso data for cancer incidence, sixty to seventy deaths from cancer are a possible figure; but it may well be many more than that.

The facts are that nobody knows precisely where the plumes blew, whether the radioactivity touched the ground, or how it was dispersed. We do know, though, that the radioactive plume was monitored 250 miles away from the plant.

But no epidemiological studies were instigated to study this exposed population. I suppose the utility and the government did not wish to know the long-term damage that the accident might have produced because then they would be up for huge expenses in compensation, and nobody wants that when they are in "business"!

Dose calculations were made by averaging the radiation dose over the totally exposed population, but some people could well have been exposed to higher risk than others because of the uneven distribution of the radioactive plume and the fact that the population is not homogeneous. The dose to the adult thyroid of iodine 131 was calculated to be 11.1 millirems, but the dose to the fetal thyroid would be ten to twenty times greater, and fetuses are many times more radiosensitive to the mutagenic effects of radiation than adults. Also, the carcinogenic effects of radioisotopes react synergistically with other carcinogens in the body to promote cancer. Therefore, depending upon the state of the environment, there very well could be an increase in various forms of cancer in the TMI area.

Dr. Gordon McLeod, former Secretary of Health in Pennsylvania, was sworn into his position twelve days before the accident. Out of town at the time of the accident, he was told about the radiation release by phone. He asked for someone in the State Health Department who knew all about the health effects of radiation release to phone him. There was no such person. He then asked for the liaison to the Bureau of Radiation Protection to call him. Again, no such person. He then requested that books on the subject be collected from the State Health Department library so he could investigate them in the morning. There were no such books because there was no library. Subsequently, Dr. McLeod initiated medical follow-up studies of the exposed population. He was also concerned about the medical consequences of the cleanup.

In fact, the accident at TMI had just begun. It took eleven years to complete the cleanup of collapsed and fragmented, extremely hot, radioactive, used fuel rods. Over 99 percent of the melted and collapsed fuel was sent to the Hanford reservation in Washington state and to the Idaho National Engineering Labs in Idaho Falls. Not only was this operation extremely difficult, but it invariably involved more contamination for nuclear workers as well as the release of more radioactivity into the Harrisburg environment. The building itself still remains extremely radioactive, much more so than a nuclear reactor at the end of forty years of operation.[1]

The one million gallons of highly radioactive water that spilled onto the floor of the reactor buildings was evaporated on site, and some of the more volatile radioactive isotopes would also have accompanied the water vapor into the atmosphere.

At the time, Dr. McLeod requested that the the State Department of Health be intimately involved in this cleanup, but as this story becomes more and more surreal, he was fired, or, as the official word came through the media, he was "requested to resign." Several people from the Pennsylvania Department of Health resigned also, and the only two departments left monitoring the operation were the State Bureau of Radiation Protection and the governor's office. Physicians seemed not to be involved in supervising or monitoring the cleanup.

A number of possible health consequences of the nuclear

power industry have been, and still are largely ignored by the NRC. Their internal report, known as the Regovin Report, states, "The NRC provided neither leadership nor management of the nation's safety program for commercial plants." In fact, according to transcripts of internal meetings during the accident, former Chairman of the NRC, Joseph Hendrie, said, "We are operating almost totally in the blind, his (Governor Thornburgh's) information is ambiguous, mine is nonexistent, and—I don't know—it's like a couple of blind men staggering around making decisions."

On another occasion, according to the transcripts, Harold Denton, who during the days following the accident was the appointed spokesperson for President Carter, said, "Yes, I think the most important thing for evacuation to get ahead of the plume is to get a start, rather than sitting here waiting to die. Even if we can't minimize the individual dose, there might still be a chance to limit the population dose." But throughout the accident, as evacuation plans were being discussed by the NRC, the radioactive plume had already moved over the population.

When Governor Thornburgh asked Chairman Hendrie, "Is there anyone in the country who has experience with the health consequences of such a release?" he replied, "Ah—not in the sense that it's been studied and understood in any real way. There were back in the days when they were doing bomb testing. They managed to give groups of soldiers and occasionally a few civilians doses in the low rem range—a subject of discussion these days, but that's about the only comparable experience that occurs to me. You are talking now about a major release, not about the small ones that have occurred thus far."

The presidentially appointed Kemmeny Commission Report (October 1979) and the Regovin Report (January 1980) both said that the state of the nuclear industry at the time of the accident was one of chaos. Accidents like TMI had almost happened twice before: once in 1974 at a Westinghouse reactor in Switzerland; and in 1977 at Toledo Edison's Davis Besse plant in Ohio. Both involved the failed PORV valve and misleading indications from the control panel, which erroneously indicated the reactor coolant system was full of water.

A brief account of the Swiss accident was submitted to the NRC, but it did not prompt notification to operators that they might be misled by their instruments if a valve stuck open. The Davis Besse accident was intensely analyzed by the NRC and was then filed away and never circulated.

Until the TMI accident, none of this was unique. The NRC received many such reports, some trivial, some important; but in either case, the information was not systematically reviewed. For the most part, these reports were summarized in a few sentences for computer listing, distributed haphazardly to various staff offices, and then simply filed away. Furthermore, there was absolutely no institutional mechanism within the NRC for developing solutions to potential dangers and ensuring that such solutions are integrated into reactor operations through design changes, new procedures, or improved training.

The Regovin Report stated, "We have found in the NRC an organization that is not so much badly managed as it is not managed at all. In our opinion, the Commission is incapable, in its present configuration, of managing a comprehensive safety program for existing nuclear power plants adequate to insure public health and safety."

At the time of the accident, there were five NRC commissioners, working ineffectively together. Most of their meetings were spent discussing isolated safety issues, personal, budget problems, administrative chores, and export licensing—very little was spent deliberating the crucial issues relating to reactor safety. This general inefficiency filtered down to reactor staff, where one of the most glaring deficiencies was a lack of any program requiring members to acquire experience in the design, construction, and operation of nuclear reactors as well as in radiation detection.

In fact, Chairman Hendrie was the only NRC commissioner with expertise in reactor engineering. Nor was the staff directly supervised on a daily basis by the commission. They were not even housed in the same state! The commissioners and their staffs worked in downtown Washington, D.C., while the NRC general staff was scattered among six office buildings in suburban Maryland. To make matters even worse, a strong we/they attitude developed on both sides. Before TMI, both the nuclear industry and the NRC had the attitude that such an accident was "not a credible event" and "would therefore never occur." This further illustrates what a potentially devastating mindset was operable at the time.

Since Three Mile Island, there has been a major shift in the thinking of most NRC members, because they now believe that an accident can happen, whereas before TMI, such notions were more a hypothesis to analyze than a real possibility. Further, the NRC has now put together rapid response teams, which are sent to sites where precursor events or serious accidents have occurred. Often they will now shut down the plant until they are satisfied that the utility is running the plant properly, and the event is not a recurring pattern or a generic issue.

The industry is now policing itself more carefully with a new Institute of Nuclear Power Operators (INPO), in an attempt to prevent a poorly operated utility causing an accident and thereby condemning the whole industry. The plant operators are now carefully supervised, and even mildly serious events are not ignored.

Nuclear power plant operators were previously trained to control normal operations but not to manage accidents. Now the training is more regular and usually includes the use of a simulator. The operators are trained not just for the stereotyped loss-of-coolant accident but how to react and investigate a range of symptoms that could develop into a more serious condition, one that had not been previously analyzed or studied.

But now, although some utility owners are still of the old school and do not believe that nuclear power plants are dangerous, most are much better prepared to handle emergencies than before TMI. Frequent exercises, conducted in concert with the federal Emergency Management Agency, test and measure the emergency preparedness of the utility and its operating staff. But some utilities are still slow to recognize deterioration of their operations and performance, and some have been recalcitrant in reporting problems and have even suppressed information.[2]

But despite this overall reassuring information, nuclear power is an unforgiving technology, and an accident could still occur by human or mechanical error. We, and future generations, still have to deal with the shockingly hazardous radioactive waste emanating from these radioactive mausoleums, which all must be closed down thirty to forty years after startup; these dates are rapidly approaching for many of the old reactors. But the NRC now wants to implement a twenty-year "life extension" for all U.S. reactors. And in their Generic Environmental Impact Statement for automatic twenty-year license extension, they evaluate a TMI or Chernobyl-level accident as "low" impact. The NRC still routinely allows plants to operate even with generic safety problems, as well as with TMI safety issues that are still unresolved.

Chapter 8

Chernobyl

A nd then there was the disaster of Chernobyl. Eight years after Three Mile Island on April 26, 1986, another dreadful but much more severe nuclear accident occurred. First detected by Swedish radiation monitors, the Soviet authorities waited forty-eight hours to tell the citizens of the neighboring town of Pripriat that they had been covered in radiation. Mikhael Gorbachev waited nineteen days before he issued an official statement confirming the Chernobyl disaster.[1]*

The Chernobyl reactor was much more primitive than the reactor at Three Mile Island. It had no containment vessel, and the control rods were made of flammable graphite instead of boron. As at Three Mile Island, the event was fueled by human

---

*In this chapter I've relied very heavily on the excellent account of the Chernobyl accident by Grigori Medvedev in the *"The Truth About Chernobyl."*[2]

incompetence. The chief engineer, N. M. Fomin, decided to perform tests while the Unit Number Four reactor at the Chernobyl complex was shut down for maintenance. He wanted to determine whether the electricity generated by the spinning turbine blades could compensate for the loss of power during a regional blackout. But a test like this must be conducted only if the emergency core-cooling system (ECCS) and the emergency reactor safety system responsible for injecting control rods into the reactor core are fully functional, and only after activation of the emergency power reduction system. And at all times the minimum number of twenty-eight to thirty control rods must be left in the reactor core to control the chain reaction. These conditions were not met. In fact, the experimental program was so poorly conceived that it required deactivating the emergency core-cooling system.

Unit Number Four began its radioactive life on December 1983. By the time of the accident, it contained two hundred metric tons of uranium dioxide, which had been fissioning long enough to produce very high concentrations of long-lived radionuclides in the core, and it was generating one thousand six hundred megawatts of thermal energy. The supervising engineer was not a specialist in nuclear reactors, having received his basic training in hydroelectric power. In fact, Gregori Medvedev claims that Yu A. Izmailov, a veteran of the Central Directorate for nuclear power in the Soviet Union, used to joke that "it was practically impossible . . . to find anyone . . . in the central directorate who knew much about reactors and nuclear physics." Medvedev said that notification of nuclear accidents was never publicized either within or without the nuclear establishment, so that nobody within the industry could learn from past mistakes. This practice has also been common in the U.S. nuclear power establishment. Two weeks before the accident an emergency button (called M. P. A. [the Russian initials for "maximum design-basis accident]) deputy chairman of the Council of Ministers of the Soviet Union said on May 6 at a press conference that radiation around the reactor measured only fifteen millirems per hour. Eight hundred tons of flammable graphite, which remained within the core, completely burned away over the next few days, spreading isotopes into the atmosphere. Inside the control room, chaos reigned as the massive explosion shattered the walls and ceiling. All the circuits shorted, and a floury white powder filled the air. People reported a strong smell of ozone and a strange metallic taste in their mouths, similar to reports from Harrisburg after the Three Mile Island accident.

Bewildered operators staggered around the crater-like ruins of the reactor. Strangely, when they reported the extent of the catastrophe to their chief, Dyatlov, he refused to believe that the reactor had been destroyed. Instead, he radioed a fallacious message to Moscow that an explosion of gas in an emergency tank of the protection and control system had blown off the roof.

Two other operators, Bryukhavanov and Fomin, repeatedly reported to Moscow that radiation levels in Pripriat were normal. It was not until an official delegation arrived from Moscow the next day that they discovered the true gravity of the situation. Young operators were exposed to radiation levels of 20,000 to 30,000 roengtens per hour, and they all died within days. Several hours after exposure they developed a "nuclear tan"—their skin became deeply brown, and they experienced uncontrollable vomiting. Some became hyperexcitable (a kind of nuclear frenzy) and then lapsed into a semicoma before they died. Others developed dreadful blistering burns, and their limbs and faces became so swollen that the skin burst. Because the patients were so contaminated the medical center also became radioactive.

Fifty fire trucks were dispatched and brave firefighters absorbed massive doses of radiation as they stood on the damaged roof pouring water into burning graphite. They were gazing into a diabolical cauldron of red and blue fire, because the 500-ton, concrete biological shield had been lifted by the explosion from the top of the reactor, landing at an angle and exposing the damaged core.

The shield later became cherry-red from the heat of the fire. Radiation readings on the roof measured 2,000 to 15,000 roengtens an hour, and seventeen firefighters were later hospitalized.

Several turbine operators also absorbed massive doses of radiation as they turned on the pump gate valves by hand to save the turbine hall. Electricians received similar huge doses. (As an aside, all four reactor units were interconnected by cables that traversed a single deacrator gallery, a design fault that could have destroyed all the reactors like Brown's Ferry.)

By the morning, about 100 people had been hospitalized.

The few dosimeters available at the plant were not equipped to measure large radiation doses. The appropriate instruments were locked in a safe. So no one had any real idea of the danger to which they were exposed.

Neither did the doctors at the medical center. It was "the blind leading the blind"; similar to the Three Mile Island tragedy. Anyway, no treatment exists for acute radiation sickness, except palliative care, so the doctors were involved in a frustrating exercise in futility.

Meanwhile, the shift foreman at the Number Three Plant had the wisdom to shut it down in the morning, thereby preventing another reactor core from melting down.

The population of 48,000 in the adjacent town of Pripriat lived life as usual on a lovely sunny day. Older children went to school, while the younger ones played in the sand or rode their bicycles. Men sunbaked on their roofs to get a tan, which they found remarkably easy that day, and women hung out their washing in the sun. The people could see the damaged reactor, and some heard the explosion during the night, but most ignored the awful reality. Some townspeople even fished in the cooling ponds and canals of the reactor right through the night, receiving doses of around three hundred rads.

But by that evening most people realized that something dreadful had happened, when they noticed a metallic taste in their mouths and smelled the burning graphite. Of course, as the graphite fires continued to burn, they filled the air with short- and long-lived isotopes, including plutonium, Americium, curium, radioactive iodine, stontium 90, and cesium 137.

On the evening of the 26th, officials decided to dump 3,000 to 4,000 tons of sand into the burning reactor from helicopters. Pilots received from twenty-eight to eighty rads as they hovered three to four minutes over the reactor while they discharged their load of sand.

One hundred and ten sorties were made on April 27, and three hundred more on April 30. By May 2nd, five thousand tons of friable material had been dumped. Obviously this procedure ejected more isotopes into the air, and months later plutonium and uranium were flushed from the blood of the pilots. They wore no respirators, nor did the planes have protective lead shielding. To assist the pilots, citizen volunteers were recruited to shovel very radioactive sand, which was located adjacent to the reactor, into bags.

Finally, at 1 P.M. on April 27, thirty-six hours after the explosion, a column of 1,100 buses moved through the streets of Pripriat to evacuate the residents. People were told to wear light clothes, to lock their doors, and to pack only enough food and money for three days. In the emergency they left all their memories and treasures behind.

The contaminated buses themselves were vectors of large quantities of radiation. They drove the people only thirty-seven miles to Ivanhov, their final destination. The hair of the citizens of Pripriat was heavily contaminated with the isotopic cocktail described above, and their clothes and bodies were also very radioactive.

By day two, the gamma radiation from iodine 131 emanating from their thyroid glands was fifty rads per hour. Shortly before the evacuation, health workers and high-school girls had distributed potassium iodide to block radioactive uptake by the thyroid gland, but by then it was too late.

Radiation at ground level in Pripriat measured fifty roentgens per hour and 50 percent of the radioactive particles were iodine 131. People ate normally, not having been warned about radioactive food intake, but they reassured themselves by drinking vodka for decontamination. Many seemed hyperactive and agitated.

All the dogs remaining in Pripriat were shot and the streets were strewn with canine corpses. Then on May 5, the people and livestock within an eighteen-mile radius of the city of Chernobyl were evacuated, and once again the dogs were destroyed. Airborn radiation increased dramatically nine days after the accident, and on May 7, ground-level radiation in Chernobyl was twenty rads per hour. Interiors of cars leaving the city measured three to five rads per hour (five hours spent in a car gave an exposure of twenty-five rads). A bunch of flowers picked in Chernobyl on May 9 measured twenty rads per hour.

Ground-level radiation in Kiev was fifteen to twenty millirads per hour, often fluctuating higher. In the first three days after the accident, the average twenty-four-hour dose received by the citizens of Kiev was 2.4 rads—two thousand times standards allowed by the World Health Organization. In the first week, one million people fled the city.

In all, half a ton, or 1,100 pounds of plutonium were released from the reactor—theoretically enough to kill every person on earth with lung cancer 1,100 times.

After the accident, soldiers were picking up chunks of nuclear fuel and graphite scattered around the reactor by hand and placing them in buckets for disposal. The chunks measured two thousand rads per hour. These people were called "liquidators," and their faces were dark brown. A total of 650,000 people participated in the cleanup, of whom 5,000 to 10,000 are known to have died so far.

On May 9, at 8.30 P.M., burning graphite in the core collapsed under the weight of five thousand tons of sand, clay, and boron carbide, releasing huge quantities of nuclear dust, which settled on Pripriat and the surrounding rich agricultural fields.[4]

Pregant women aborted who had been exposed to radiation in contaminated areas. The nuclear accident at Chernobyl will go on forever and will linger in the genetic material of future generations for the rest of time.

Large areas of the "breadbasket" of the Ukraine and Byelorussia are heavily contaminated and will remain so for thousands of years. Five million people still live in these areas, one quarter of whom are children.

The radiation also fell heavily over areas in the countries of Austria, Bulgaria, Czechoslovakia, Finland, France, East and West Germany, Hungary, Italy, Norway, Poland, Romania, Sweden, Switzerland, Turkey, Britain, the Baltic States, and Yugoslavia. It spread around the globe, and small amounts fell across the United States, Canada, and all countries in the Northern Hemisphere.[5]

There was great distress in Europe about radioactive food, water, and air, soon after the accident, but within a year the concern abated. It is obvious that most governments in the involved countries either do not understand that strontium 90 and cesium 137 last for six hundred years, and that plutonium remains radioactive for five hundred thousand years—or they choose to ignore this extremely difficult situation either for economic or political reasons.

These long-lived isotopes concentrate thousands of times in the food chain, and people will be eating radioactive food for the rest of their lives; the lives of their offspring will also be affected. (There are signs posted in Bavarian forests warning people not to eat the mushrooms, which are apparently efficient concentrators of radioactive isotopes.)

John Gofman M.D., an eminent radiobiologist who used to work for the U.S. Atomic Energy Commission, estimates that the total number of fatal cancers caused by Chernobyl will be between 140,000 and 450,000, with an equal number of nonfatal cancers (some cancers are curable). This is almost one million cases of cancer.[6] The Department of Energy in the United States, however, which oversees and supports the U.S. nuclear power program, estimates that only 17,400 fatal cancers will be induced over the next fifty years, an artificial cutoff time.[7,8]

Note that Gofman's estimate is only related to cesium 134 and cesium 137 induced cancers, because cesium is the dominant isotope in the Chernobyl fall out. His estimates do not include iodine 131 related thyroid tumors; 19,500 leukemias—malignancies induced by other short- and long-lived isotopes, or the number of genetic or teratogenic defects to be expected.[9]

Each 1,000-megawatt reactor produces nearly 4 million curies of cesium per year. Extrapolating from this figure, if 400 reactors operated for twenty-five years at 99.9 percent perfect cesium containment, cesium loss from startup to burial would be equivalent to sixteen Chernobyl accidents every twenty-five years.[10]

Gofman says "the Chernobyl accident dismayed the promoters of nuclear power in virtually every country of the globe," and he implies that it is in the best interest of these organizations to downplay the ongoing medical consequences.[11] The Bush presidency and the DOE promoted the construction of

175 new reactors in the United States over the next forty years.[12]

ECOLOGICAL CONSEQUENCES OF CHERNOBYL

Radioactive dust is transported by wind currents, and isotopes are conveyed by water currents, migrating birds, and wildlife.[13] But human activities also play an important role in the distribution of man-made radiation. Obviously, fresh and processed food exported from certain regions of Europe is radioactive. Australia imports many varieties of processed food, including jam and pickles from Poland, cheese from Germany and Scandinavia, and tomatoes, olive oil, and pasta from Italy. After I heard that herbs from Turkey were highly radioactive, I called the inspector in charge of testing imported food for radiation in Australia, and I said "How do you test for radiation?" He said, "We do random spot checks," meaning the computer picks out certain batches of food to be tested. This method is clearly imperfect because batches of radioactive food could be missed. When I asked, "What do you do when you find radioactive food?" he blithely said, "We dilute it with non-radioactive food." As we have seen in this book, the solution to pollution by dilution is fallacious when it comes to radiation. Australia grows the cleanest food in the world—we should be exporting our nonradioactive food to a polluted world. Since Chernobyl, researchers in Quebec have found that caribou meat eaten by the indigenous population contains radioactive cesium.[14] A sheep farmer in Wales was told to leave his land because his lambs were radioactive. When he asked "for how long?" the authorities answered, "About 100 years."[15]

The dose of Chernobyl radioiodine to infant thyroid glands in Scotland was almost equivalent to forty-five chest X rays.

(Iodine 131 remains radioactive for only six weeks.)[16] *The Economist* for May 10, 1986, stated that the reaction to Chernobyl in Belgium and France was quite calm, mainly because both countries generate half their electricity from nuclear power, and the dangers were therefore underplayed publicly. French officials did not inform farmers about the concentrations of radioactive fall out on their properties. Britain also put public perception before public health, because the U.K. nuclear industry was made insecure by Chernobyl. The head of the Radiological Protection Board, a government body in Britain, said, "there will be cesium in the silage this winter and there will be cesium in milk and meat next year. I would not like to predict the effect on people." This statement was not widely reported. On June 26, 1986, the *New Scientist,* four weeks before lamb sales were banned, reported that levels of cesium in the liver and muscles of Cumbrian lamb were far above levels that should trigger precautionary measures. The biological half-life of cesium is 130 to 365 days, meaning that it stays in the body for decades. And as its radioactive life is 600 years, it remains in the soil to concentrate and reconcentrate in food over many human lifetimes. Have the English or European people been told these facts, or warned about their food? The answer is no. This is a huge public health scandal because the authorities know, but won't protect the public.[17]

Officials in the Ukraine buried 400 tons of radioactive beef after Chernobyl, and another 920 tons was buried in June 1992. In 1990, it was reported from Moscow that butter, meat, milk, grain, and potatoes were still being grown in highly contaminated areas of Byelorussia and distributed all over the Confederation of Independent States. One-third of Byelorussia is contaminated, and one-fifth of its arable land is "dead." In the villages of Lomachi and Tulgorvichi, 5.4 curies of plutonium per square kilometer contaminates the soil (previous residual level from atomic testing was 0.1 curies).[18]

Estimates of $26 billion are touted for relocation of the affected population in the Confederation of Independent States, and for the provision of clean water and adequate medical care.[19] But by September 1991, the United Nations had allocated only $6 million for this project.[20] Yet in 1990 the United States and the United Nations spent $90 billion to finance the Gulf massacre. Money is always available for war but rarely for compassionate humanitarian purposes.

NUCLEAR POWER POST-CHERNOBYL IN THE CONFEDERATION OF INDEPENDENT STATES

When I visited a nuclear reactor near Moscow in Russia in 1979, it was obvious that an accident was waiting to happen. The secondary cooling water from the reactor was used to heat the houses in the adjacent town. The external concrete shell of the reactor was cracking, there was no containment vessel or ECCS, and the health officer categorically denied that any radiation escaped during routine operations. When I challenged him he left the room in a huff; we all drank a lot of vodka that night to placate him.

But six years after Chernobyl, fifteen similar light-water, graphite-moderated reactors continue to operate, producing half the nuclear power in the CIS. Ten old, Soviet-designed, pressurized water reactors operate in Czechoslovakia and Bulgaria. And in March 1992, a Chernobyl-type reactor adjacent to St. Petersburg experienced a serious radioactive leak, the fourth accident involving such a reactor in a year. These reactors are huge, have no ECCS, and the graphite burns like a matchstick; but unless they blow up or breakdown, they probably will operate into the next century, because the only viable alternative in these countries is dirty brown coal.[21]

Meanwhile, the morale of operators at Chernobyl-like reactors is declining. The plants are inadequately financed, and they may soon be retired. To aggravate the situation, power is sold for only 20 percent of production costs; control of the reactors has been diversified from centralized Moscow administration to the various independent states; and accident investigation is still focused on fining individual operators, rather than eliminating the engineering or procedural problems.[22] There are forty-five reactors in the CIS[23] although in May 1992, Ukrainian authorities shut down the last two working reactors at Chernobyl.[24] Despite the ongoing tragedy of Chernobyl, on March 26, 1992, the First Deputy Prime Minister of Russia, Yegor Gaidar, signed an order to resume construction of a number of new reactors and to increase the capacity of the existing ones.[25]

A confidential report issued by the World Bank and the International Energy Agency in Paris in June 1993 says that 25 of the most dangerous nuclear reactors in six nations of the former Soviet Union could be closed in the next few years and replaced with cheaper gas fired plants. But Russia continues to insist on nuclear energy, despite the danger to the whole of Europe, because it wants to reserve its natural gas supplies for export in order to earn hard currency![26]

In the civilian nuclear industry in 1991, there were 165 various accidents at nuclear reactors.[27]

OTHER RADIOACTIVE DISASTERS IN THE CONFEDERATION OF INDEPENDENT STATES

Since Glasnost and the collapse of communism, the truth is finally being told about previous unannounced nuclear disasters. But as I recount these environmental tragedies, remember that they mirror a similar situation within the United States. Both countries opened a Pandora's box when they decided to ignore Einstein's warning that "the splitting of the atom changed everything save man's mode of thinking, thus we drift towards unparalleled catastrophe," and to confront each other with a vast array of nuclear weapons and produce electrical generation by nuclear power.

Now it's too late. Vast areas of both land masses and numerous water bodies are contaminated forever with radioactive waste scattered at random by governments and industry too preoccupied with power, money, the cold war, and the fascinatingly dangerous challenge of the "hard energy" option.

The United States, which, we will see, has huge nuclear waste problems of its own, launched a $20 million plan to reduce the risk of nuclear accidents in Russia and the Ukraine. This money would provide centers to train personal in safety procedures. But inherent danger resides in each of the 430 reactors on earth, which could melt down from technical or human error. The director of safety at the International Atomic Energy Commission says we can expect a major accident every ten years,[28] and America's own NRC predicts a 50 percent chance of a meltdown larger than Chernobyl by the year 2000.[29] To add fuel to the nuclear fire, a former managing editor of *Nuclear Fuels and Nucleonics Week,* a trade journal of the nuclear industry, said, "[The industry] downplays environmental considerations."[30]

There is deep concern that former Soviet nuclear scientists who are now unemployed will defect to Middle Eastern or other unstable countries to help them develop nuclear weapons. To avoid this happening, Japan, the European Economic Community, and Russia have jointly established a fund of $75 million to create an International Science and Technology Center in the former Soviet Union, which would use the talents of two to three thousand scientists for universities, for commercial nuclear research, or for foreign investors or governments willing to pay.[31] But this may not entice all the scientists to abide by the wishes of these advanced nuclear nations.

The nuclear cat is already out of the bag. Iran has secretly negotiated to buy nuclear weapons from Kazakhstan, which they may deploy on Chinese-made Silkworm missiles,[32] and China bought large numbers of weapons from the bankrupt CIS, some of which possess nuclear capabilities.[33] In the modern world, no country has a monopoly on nuclear secrets.

The radioactive mess in the CIS is almost too terrible to contemplate. Former Soviet officials secretly dumped sixteen nuclear reactors from submarines and an icebreaker into the Kara and Barents Seas, near the Arctic islands of Novaya Zemlya. About half of the reactors still contained their irradiated fuel. A complete submarine that experienced a major nuclear accident was also scuttled, together with radioactive components from other submarines. Three of the reactors are from the icebreaker *Lenin,* which had experienced a serious accident. Two reactors were dumped into the Sea of Japan. More than seventeen thousand containers of high-level waste were discarded over the years in shallow seas around the islands, and those that failed to sink were shot and filled with holes by the Soviet navy. The Russians say they have almost no direct data about whether these containers are intact, corroding, or breached. Waste water from civilian and naval reac-

ing technology that emits no sulphur dioxide, nitrous oxides, or greenhouse gases. In fact, nuclear energy helps "reduce" airborn pollutants in the U.S. by over 19,000 tons per day. That's because the 111 nuclear plants now operating in this country don't burn anything to generate electricity. The air we breathe is cleaner because of nuclear energy. But we need more nuclear plants. Because the more plants we have, the more energy we'll have for the future of your planet."[4] Their slogan is "Nuclear energy means cleaner air."

Let's dissect the above assumptions. It was estimated in 1974 in a Friends of the Earth paper, that a nuclear reactor must operate at full capacity for ten years to repay its energy debt incurred by uranium mining, enrichment, fuel fabrication, steel and zirconium manufacture, and plant construction (this does not include decommissioning). Add to this eight to ten years for plant construction and fuel loading and reloading. It would therefore take approximately eighteen years for one net calorie of energy to be generated for societal consumption.[5] These operations generate large quantities of carbon dioxide gas so the nuclear energy advertisements are a lie. Nuclear energy adds to the greenhouse effect.

Even the American Medical Association (AMA) is an apologist for nuclear power. A 1989 paper entitled "Medical Perspective On Nuclear Power," published in the Journal of the American Medical Society, endorsed the nuclear industry. This paper recommended that nuclear power generation is acceptably safe in the United States, that worker exposure to radiation has decreased in the last decade, that each state should be responsible for storage of radioactive waste, and that physicians should be responsible for advising and treating the public in the event of a nuclear accident. The problem with the latter statement is that there is no effective treatment for irradiated patients, and decontamination of human bodies is practically impossible.[5] The Council On Scientific Affairs, which formulated the AMA document, was staffed by staunch advocates and employees of the nuclear industry.

In fact, most of the doctors and nurses at the Harrisburg Hospital fled following the Three Mile Island meltdown, leaving the patients to fend for themselves. When I briefed the Harrisburg physicians after the TMI accident, their knowledge about the medical consequences of alpha, beta, and gamma emitters emanating from nuclear reactors was rudimentary, but intuitively they had realized that the radiation release was extremely dangerous. President Bush's new energy plan called for the construction of approximately 175 "advanced" nuclear reactors over the next 37 years.[6] This plan assumes that the present 110 reactors will continue operation, although most will be older than their designed operational lifetime and will therefore be that much more radiologically dangerous.

Although the presidency changed in 1992 it is important for us to understand what the nuclear energy industry has in store for us if it can get its way. And on a cautionary note, I would warn you that Vice President Al Gore takes a pronuclear line in his ecological book, *Earth in the Balance.*[7] President Clinton, in his energy budget, has in fact cut research funding for the advanced generation of the high-temperature, gas-cooled and the PRISM breeder reactors, but this funding could be restored if the Congress succumbs to intense lobbying by the nuclear industry, or if a Republican president is elected in 1996. As I have previously warned, there is an intense advertising campaign in effect by the nuclear industry, and they therefore must have plans to continue constructing reactors.

Three typical "advanced" nuclear reactor designs have been analyzed by MHB associates for the Union Of Concerned Scientists.[8] One of these reactors is a Westinghouse-designed, AP-600, pressurized water reactor of 600-megawatts. The Design Certification Program for this reactor is partially funded by the Department of Energy. The AP-600 is a simplified system compared with existing reactor designs, requiring fewer pumps, valves, pipes, and control cables. These plants will be semi-mass-produced, or prefabricated, in order to lower the costs of design and construction, and to decrease construction time.[9] Some of the safety features, which are described as "passive," rely on gravity and nitrogen gas pressure. But the instruments that are essential to facilitate or activate the safety systems are not passively operated.[10]

General Atomics is designing a Modular Gas, High-Temperature, Gas-Cooled Reactor—MHTGR. The standard plant will consist of four helium-cooled, graphite-moderated reactors, producing a net output of 538 megawatts. Designed for a lifetime of forty years,[11] it will be located below ground, although the two steam turbine generators and the reactor cavity cooling system will be above ground, vulnerable to sabotage, and fires.[12]

The fuel will consist of millions of microspheres of enriched uranium oxycarbide coated with two layers of pyrolytic carbon and one layer of silicon carbon, together with thorium oxide microspheres, sealed in graphite blocks. Each reactor will contain about 10 billion fuel kernels. The coating, like the conventional zirconium fuel rods, will prevent release of fission products, except when temperatures exceed 1,600 degrees centigrade, when the coating fails (at the same temperature as zirconium). The original MHTGR design did not have a containment vessel.[13]

Although this particular reactor has never been built, the Department of Energy (DOE) decided that a prototype demonstration plant was unnecessary, although the Nuclear Regulatory Commission (NRC) staff disagreed. The DOE recommended instead, that a four-reactor complex be constructed at the Idaho National Engineering Lab, to produce tritium for new nuclear weapons and to update old weapons, although the cold war is over. DOE planned to start up the first module in 1998, but President Clinton has now canceled this military program.[14] This particular reactor design has the potential for large nuclear accidents that could be triggered by graphite fires (as in Chernobyl), steam generator tube ruptures, and control rod ejection that would cause graphite damage and reactivity insertion—an increase in the flux of neutrons in the reactor core.[15]

The third and most dangerous "advanced" reactor planned by the nuclear industry is the PRISM, or Power Reactor Inherently Safe Module. This is a plutonium breeder reactor. PRISM will be cooled by liquid sodium, which is highly reactive and burns when exposed to air. It reacts chemically with concrete, it explodes on contact with water, and if it boils it can cause a nuclear explosion in the plutonium fuel.[16]

Despite these enormous risks, a containment vessel is not part of the PRISM design, although the NRC staff have determined that a meltdown is possible.[17]

Past U.S. experience with breeders has been a disaster. The Fermi breeder reactor operated for 6 years before it partially melted down near Detroit in October 1966. Several experimental breeders operated for only relatively short time spans.[18]

Thus we enter the plutonium economy, a diabolical pact with the devil, where for each pound of plutonium fissioned, 1.2 pounds will be made.[19] Plutonium fuel will be separated from spent fuel from breeders, and also "mined" or reprocessed from the vast quantities of spent fuel rods currently resting in swimming pools at old light-water reactors.[20] It will then become available for accidental release, spontaneous ignition, criticality accidents, lung cancer, and bioconcentration in the food chain. At the time of writing, Japan has launched a ship containing 1.7 tons of plutonium, reprocessed in France from its spent fuel. The ship will travel thousands of miles, probably through the Southern Ocean south of Australia on its way to Japan. The United States government has approved the shipment, and the Australian government has not objected. Should there be an accident, depending upon the wind currents, much of the Australian continent and its people could be contaminated with plutonium. Thus Japan enters the plutonium economy a little ahead of the United States of America.[21]

When I wrote the original edition of *Nuclear Madness* in 1978, I feared but never really thought that the nuclear priesthood would ever actually indulge in plutonium madness. But despite the dire warnings of TMI and Chernobyl, it seems to have been irresistible.

General Electric, Bechtel Power Corporation, and others are designing PRISM, and the research and development is being conducted by nuclear weapons labs—Argonne National Lab, Energy Technology Engineering Center, Hanford Engineering Development Lab, and Oak Ridge National Lab. One would wonder why these labs are so involved, until we realize that transmutation of disguarded plutonium from dismantled nuclear weapons is the new order of the day.

Long-term plans call for the disposal of the plutonium derived from these decommissioned former Soviet Union and U.S. weapons. The plutonium will be placed in reactors and fissioned and converted or "transmuted" to shorter-lived, radioactive isotopes; e.g., from plutonium 239, with a total life of 500,000 years to strontium 90 and cesium 137, with radioactive lives of six hundred years.[22] Sounds good, but I wonder if eighteen generations from now, the people will appreciate the carcinogenic and mutagenic effects of ubiquitous strontium 90 in their bones, and cesium 137 in their muscles and reproductive organs. General Atomics agreed in April 1993 with the Russian Ministry of Atomic Energy to build a reactor in Russia which would fission plutonium from decommissioned weapons for fuel. It will require $100 million from the U.S. government over the next five years, costing a total of $1.5 billion. The reactor will be a modular HTRG model.[23]

To return to the PRISM reactors. The total PRISM complex

will consist of blocks of three modular reactors connected to a single turbine generator, with a total of three blocks (nine reactors). Each small modular reactor will produce 137 megawatts, so there will be a net output of 1,245 megawatts. The reactors and generator will be located underground, and the nine reactor modules will be operated from only one control room,[24] which, at full power, will be fully automated, almost removing the need for human operators.[25] The PRISM reactors do not have an emergency core-cooling system of the type used in light-water reactors.

A reprocessing plant will be built on site to extract the plutonium, which will be reused over and again in the breeders. Fuel fabrication and waste treatment plants will also be on site. And despite the fact that this is new, dangerous, and untried technology so far, the NRC does not require the construction of a prototype test plant.[26]

The designers of both the MHTRG and PRISM reactors assert that operators are not necessary for safety, and that the control rooms themselves do not perform a safety function.[27] In fact, the four-module MHTGR will be staffed by only one senior reactor operator, two operators in the control room, and five roving operators[28] (former NRC regulations required two senior operators and three other operators to run a two-reactor control room). MHB Technical Associates contends that because there is virtually no experience with these new reactors, they cannot be presumed to meet all necessary safety requirements, and that the MHTRG should therefore be equipped with a conventional safety related control room.[29]

In the PRISM plan, three operators control nine reactors from one control room. So one operator could simultaneously be faced with a catastrophic situation triggered by loss of off-site power in a unit at full power, in another shut down for refueling, and in one in startup mode. And the MHTRG reactors have similar problems.[30]

The DOE also plans to classify many systems such as fire

alarm and detection systems, radiation monitoring systems, steam and water dump systems, and a seismic monitoring system as non-safety-related; while they are unarguably safety-related.[31]

Because many of the mechanisms within these reactors are operated by the force of gravity or by gas pressure, they are classified as passive safety systems, and because of this new design element, the vendors have classified their new dangerous reactors as "inherently safe."[32]

The rationale behind the scaled-down safety requirements on PRISM and MHTGR is obviously money. Fewer operators, no ECCS, no containment vessels, and reliance on "passive" safety mechanisms make the reactors cheaper and easier to build.

But the MHB team states that "as a general proposition, there is nothing inherently safe about a nuclear reactor." Regardless of the construction, design, operation, or management, there is always some mechanism that would render the reactor dangerous. And precisely because there is no operating experience available for these hypothetical paper designed reactors, all notions of safety and security at this stage are mere fabrications.

The reactor designers are so confident about the adequacy of their safety systems that they have even eradicated public emergency evacuation and protection plans. But the MHB team says that these new designs are vulnerable to multiple module or reactor accidents.

Even the NRC is so insecure about these aberrant claims of safety, that they expect "at least the same degree of protection from the public and the environment that is required of the present generation of light water reactors." Under present laws the "advanced" nuclear reactors would not be given a license. In this context in 1985, the NRC estimated the chance of a major reactor accident to be 45 percent during the years 1985 to 2005.[33]

Nuclear waste storage has not even been addressed by the vendors, nor has the very real potential of plutonium theft and weapons proliferation.[34] And at this stage, it is almost impossible to obtain detailed designs of these advanced reactors because they are classified by the designers. Even probability risk assessments are withheld from public scrutiny.[35] If this proprietorial information is not made public, it will be impossible for the public and the politicans to monitor this dangerous new enterprise.

Yet despite these very serious problems, the NRC has proposed streamlining licensing procedures virtually to remove public input relating to the new generation of reactors. To this end, it changed the licensing procedure under a new generic regulation, 10 CFR, Part 52.[36] The new rules are as follows:

SITE PERMITS

A utility can now obtain a permit to construct a reactor, or group of reactors on a specific site, twenty to forty years before construction begins. It may then take another ten years for construction time, followed by forty to fifty years of reactor operation. This means that a site can be chosen one hundred years before the reactor closes. The decision to build the reactor cannot be challenged in the interim even if new seismic data is discovered that is incompatible with the original reactor design, if the surrounding population density increases, or if these reactors are found to be more vulnerable than expected to fire, flood, and tornadoes. And the public will have no right to intervene under this new regulation.[37]

DESIGN CERTIFICATION

A design for a new reactor can be certified by the NRC thirty years before it is built, despite the fact that much could be learned about the reactor in the interim that may be deleterious to its proper functioning. But, as MHB points out, most design problems are only discovered years after the plant goes into operation. This new law effectively excludes public access to relevant data on plant design, and the discovery of operational problems before construction of a possibly flawed design.[38]

COMBINED LICENSING

This is a single procedure that includes authorizing a construction permit and an operating license in one fell swoop. Previously, these were separate procedures that were open to public scrutiny and input. In the past, some brave people have spent years of their lives dedicated to these sorts of hearings. But the public will now be virtually excluded from intervention, because with the new rules a person can only file an objection sixty days before the plant goes critical.

The objective information necessary for intervention must be of such technical sophistication that only a mole within the industry could be well enough informed.

And even if a petition is granted by the NRC at this late stage, fuel loading and operation will go ahead unless the petition is put into immediate effect. Remember, too, that neither site reliability nor reactor design can be questioned at this juncture.[39]

After intense lobbying by the nuclear industry, and active opposition from a broad coalition of environmental and public interest groups, one-step licensing was passed by the Congress on May 27, 1992.[40]

The nuclear industry plans to construct the nuclear modules in thirty to forty months instead of the usual one hundred and five months for a light-water reactor, and plants will be permitted to operate for forty to fifty years instead of thirty years for light-water reactors.[41]

So now we face a situation where new, untested, potentially unsafe reactors may be built over the next thirty-five years, with even fewer licensing safeguards than in the past—at a time when nuclear waste pollutes the earth, and millions are destined to die of radiation-induced disease; all following a theory fabricated by the nuclear priesthood supposedly to mitigate the greenhouse effect—which, in fact, nuclear power cannot do.

When will we be free of their absolute control over our lives? What will stop them? A meltdown at Indian Point forty miles from the center of Manhattan?

During the Reagan years, 1980 to 1988, funding for alternate energy research was cut almost to zero at DOE. George Bush increased funding for solar power from 0 to 5 percent, while increasing nuclear funding by 40 percent.[42] So virtually no work has been done by the DOE over the last thirteen years on clean alternate energy, and all funding and research has been spent devising this new generation of standardized "inherently" safe nuclear reactors, despite the lessons of TMI and Chernobyl.

Meanwhile, U.S. utilities are becoming disenchanted with nuclear power, because they realize that it is not cost-effective or competitive in a free market. It is very possible that the transnational nuclear vendors, GE, Westinghouse, Bechtel, et al., will be unable to sell their atomic wares in the United States Over the last five years, nuclear energy has become more

expensive than coal or oil, and utilities have been subsidizing
the installation of energy-efficient light globes, motors, and
other devices, because they have learned that it was less costly
to save energy than to make it.[43]

But because of the stagnant demand and public opposition
in the United States, western nuclear power companies are
heading toward Asia. Woodrow Wilson, manager of the advanced reactor commercial programs for General Electric, the
world's biggest power generating equipment supplier, said in
June 1993, "The Asia Pacific region has by far the greatest
market potential in the next decade." South Korea is planning
eighteen reactors over the next thirteen years, Taiwan has
plans for a fourth reactor, Japan already has forty-two commercial reactors with ten more under construction, and other
lucrative markets include Indonesia, Thailand, Malaysia, and
the Philippines. The World Bank estimates Asia's energy needs
will double in the 1990s. And China, which has just one reactor
with another one about to go on line, is planning at least two
more.[44]

A recent conference at the Massachusetts Institute of Technology, organized by the Nuclear Engineering Department,
recognized that the biggest problem for the nuclear industry
was the sticky issue of public trust, which they had never
actually tackled. Instead, they have always substituted public
relations to gain acceptability.[45] But the public doesn't believe
them. As one witness from TMI said, "It's like a child having
a grenade in its hand and you lay it out there for him to play
with. Sooner or later he'll figure out how to pull the pin and
blow his brains out. That's what they're doing down there with
nuclear. Just playing with it."[46]

The American Nuclear Society holds a series of week-long
workshops for school teachers, funded in part by a grant from
the DOE, co-sponsored by the American Nuclear Science
Teachers Association. The teachers are shown how to teach

children about nuclear energy, and one teacher said at the end of the course, "I feel better about the safety of nuclear power plants. I think steps should be made to reduce high-level waste (as in breeder reactors) because I see this as the weak link in the fuel cycle." They had obviously been taught the "benefits" of plutonium transmutation in fast breeders.[47]

Penn State University, through its Nuclear Engineering Department, sponsors programs to complement the Philadelphia Electric Company's public outreach for the Peach Bottom and Limerick nuclear reactors. These include adult education, school and scout programs (grades K through 12), and teacher programs. In 1991, the university provided 402 school presentations to ten thousand students, 68 programs to two thousand adults, and a summer camp for boy scouts.

It is immoral in my view to allow commercial enterprises to influence our children with biased data, particularly when the long-term side effects of the industry are teratogenesis, leukemia, cancer, and genetic disease. Florida Power and Light has opened an energy information center called Energy Encounter at the St. Lucie nuclear reactor located in an area surrounded by a growing population and numerous schools. There are fifty of these centers in the United States. These facilities propagandize about one million people each year. Florida Power and Light president said, "We are demystifying nuclear power plants, helping schools teach curriculum related to energy." Their promotional brochure is distributed at the Florida turnpike service plazas, chambers of commerce, and local restaurants and motels.

Power plant employees staff the propaganda center. Children are taught how to pack fuel rods with simulated uranium pellets, and to build nuclear reactor models. In recognition of its excellent public relations record, the center was awarded a Silver Anvil prize by the Public Relations Society of America.[48]

Carl Goldstein of the U.S. Council for Energy Awareness

says that the nuclear industry needs to calm the public. In 1991, the council conducted media tours by nuclear specialists to seventy cities, nuclear energy forums in eighteen cities, and wrote two hundred letters to the editors of local newspapers and 112 op-ed articles. Emphasis is placed on the connection between nuclear power and nuclear medicine—an argument that is irrelevant, because only very short-lived isotopes are used in nuclear medicine and they can all be made in a cyclo-tron. I recently debated one of these "specialists" on TV. When I accused him of being a "public relations" person for the nuclear industry and the interviewer backed me up, he became virtually speechless.

In Canada, nuclear power ads are placed in environmental magazines such as *Canadian Geographic* and *Greening Of Canada,* and commercials are broadcast during baseball and hockey games, talk shows, and day-time serials on TV. Earl Taylor of Cambridge Reports/Research International advises utilities to be "green," to institute recycling and tree planting programs in the local area. In Japan, a nuclear company opened a bird park and a visitors center on the reactor grounds.[49]

When will we be free of the nuclear industry's absolute control over our lives? What will stop them? A meltdown at Indian Point, forty miles from the center of Manhattan, trap-ping millions of people in a radioactive hell, unable to escape, dying within forty-eight hours of acute radiation illness? Such an event is not unlikely according to the NRC, because this reactor is plagued with safety problems.[50]

Chapter 10

The Cold War Ends and The Hot War Starts

If the Confederation of Independent States is a radioactive mess, what about the United States? At the same time that former Soviet officials began revealing their radioactive secrets, the U.S. Department of Energy had been forced through the Freedom of Information Act to reveal its terrible secrets, formerly obscured behind the classification of "national security." I would venture to suggest that this is the biggest cover-up in human history.

In October 1988, the *New York Times* published twenty full-page stories revealing shocking contamination at fifteen major DOE weapons-producing facilities in thirteen states.[1] By 1989, it was discovered that there are 3,200 sites in one hundred locations owned by the DOE that have contaminated soil, ground water, or both; but actually there are 45,000 potentially radioactive sites around the United States, and twenty thousand of them are government owned.[2]

Since World War II, these weapons factories have made

more than seventy thousand nuclear weapons.[3] These U.S. weapons facilities, which currently employ one hundred thousand people, have always been run by private contractors under government contract. Over the past fifty years they have employed over six hundred thousand people.[4] Their purpose was to produce weapons as fast as possible, and environmental and health concerns were almost totally disregarded. There was no public scrutiny because the cold war required absolute secrecy.

At every site throughout the land from Savannah River, South Carolina, to Hanford, Washington, a most ghastly mess was created and left. Soil, underground water, and surface water in lakes, creeks, rivers, and lagoons were extensively contaminated, while wildlife and humans were contaminated randomly and extensively with radioactive isotopes, heavy metals, and poisonous organic compounds.[5]

Yet, despite current widespread concern, and billions of dollars dedicated to cleanup, there is still no real comprehension of the amount of material dumped at each site. There is not even an effective method to track and map the extent of the radioactive contamination, including the migratory plumes in ground water, soil, and rivers.[6]

No meaningful effort was ever exerted adequately to assess the short- or long-term health consequences among the contaminated populations. In fact, no data were collected, and no retrospective or controlled epidemiological studies were done. In most cases the populations have never been officially informed.

Even now, after all the hue and cry, the research and money allocated by Congress is focused solely on locating and quantifying the waste, and on technological methods to fix it, but not to care for the distressed contaminated populations who are developing malignancies.

It is also surprising to realize, in light of the current disaster, that no appropriate technologies have yet been devised to cleanup, remove, or to stabilize the waste.[7]

DOE has promised to "cleanup" all the sites in thirty years, but a Congressional Office of Technological Assessment study is very sceptical of this estimate because "it is not based on meaningful estimates of the work to be done or the level of cleanup to be accomplished at the end of that time."[8] DOE does not know either how much waste there is or where it is, because few if any adequate records of weapons complex radioactive waste disposal were kept over the years.

The only way to describe the production of nuclear weapons is environmental vandalism. None of the corporations involved seemed to give a damn about dumping poisons into the earth, water, or air, because they were so obsessed with their money making, bomb making, and defeating the Russians.

Of course cleanup of long-lived radiation is a euphemism— once it is cleaned up, where will it be put? Reburied in someone else's back yard? And since the original site can never be adequately decontaminated, we end up with two radioactive sites, not one.

Relevant government departments, including the DOE, have always maintained that the contamination poses no "imminent threat," which is not and never has been substantiated by scientific evidence. The factory grounds themselves served as waste disposal sites; radioactive waste and toxic by-products were buried into the soil beside the buildings, dumped into nearby available water, and released into the air at random.[9]

Among the main corporations and universities involved in this nuclear madness are AT&T—the friendly telecommunications network that runs Sandia Labs; Westinghouse, which also makes refrigerators, stoves, and irons; GE, which "brings good things to life"; and the University of California, which, by rights, should be teaching young people how to survive in a safe, clean world.

I will now briefly describe the function of each facility, the radioactive and chemical contamination they have provided to their workers and surrounding communities, and the scant data available on health effects.

Lawence Livermore National Laboratory was established in 1952, near Livermore California. Both this laboratory and the Los Alamos Labs established in 1943 in New Mexico, have been the dynamos of the massive U.S. nuclear weapons buildup. Sponsored by the University of California and staffed by thousands of brilliant scientists, they have conducted research and development of all U.S. nuclear weapons and have tested more than 730 above and below ground in the Nevada desert.

Lawrence Livermore also led the research and development into the Star Wars technology; since President Reagan's Star Wars speech in 1984, billions of dollars have been poured into fruitless plans to launch nuclear weapons into space that could produce X-ray laser beams to shoot down hostile Soviet missiles.[10] President Clinton's Secretary of Defense, Les Aspin, has renamed the Star Wars Program, from the Strategic Defense Initiative to the Ballistic Missile Defense Organization, but the budget of $3.8 billion remains unchanged. The more things change, the more they remain the same.[11] Although the cold war is over and all new work on nuclear weapons has been canceled since Jan 1992,[12] the 1993 DOE budget contains funding requests sponsored by the labs for "exciting" new weapons, including advanced earth penetrators for deeply buried targets; low-yield, electromagnetic radiation weapons; hypervelocity aircraft-delivered weapons, and the strategic flow power radio frequency weapon.[13] The Los Alamos Labs are also designing micronukes—ten tons of TNT-equivalent; mininukes—100 tons of TNT equivalent; and tinynukes—

1,000 tons of TNT equivalent. They say that very low-yield nuclear weapons could be very effective and credible counters against future third-world nuclear threats.[14]

The weapons labs are so attached to their hydrogen bombs that at a January 1992 Los Alamos conference, the Star Wars scientists, including Edward Teller and his protégé, Lowell Wood from Lawrence Livermore, called for a fleet of 1,200 powerful missiles to transport the entire world's nuclear arsenals to shoot down asteroids in space. Wood could not contain his excitement at one point, calling out "nukes forever," and Teller called for a bomb ten thousand times larger than any bomb previously built.[15] This account sounds far-fetched, but it gives some insight into the psyches of some of the leading bomb makers.

In light of the push for "advanced" nuclear reactors, Livermore is presently conducting demonstrations of a new Atomic Vapor Laser Isotope Separator technology (AVLIS), designed to produce fuel for civilian reactors.[16]

CONTAMINATION

Since 1960, Lawrence Livermore has released 750,000 curies of tritium into the atmosphere. Tritium is radioactive hydrogen with a half-life of 12.3 years (and a hazardous life of 240 years). It can incorporate into the DNA molecule, and cause mutations or chromosomal damage. The lab plans to resume emmitting tritium in the near future, according to a March 12, 1992, draft environmental impact statement.[17]

No epidemiological studies have been performed on the surrounding populations. However, the incidence of testicular cancer; prostate, salivary gland, colon, breast, and bladder cancers; and melanomas and non-Hodgkins lymphoma is elevated among past and present employees at Lawrence Livermore Lab.[18] These workers commonly handle various radioactive and toxic chemicals, including plutonium. Tritium and volatile hydrocarbons have been found in ground water and soil.[19]

I refer you to a list of toxic chemicals and radionuclides commonly used and then dumped as waste at weapons facilities.[20]

Los Alamos National Laboratory is located near Albuquerque, New Mexico.

Table 3 indicates the list of hazardous substances released into the environment since the lab opened in 1942.[21] It is important to understand that the organic and inorganic compounds appearing in this table are used to extract plutonium and uranium from irradiated fuel and from old bomb cores. When these chemicals are mixed with the highly carcinogenic radioactive isotopes in the waste stream they facilitate the migration of those isotopes through water and soil. Most waste sites at these weapons facilities are contaminated with mixed chemical and radioactive waste.[22]

Los Alamos produces 1.5 metric tons of plutonium a year,[23] enough to fuel fifty to one hundred nuclear weapons.[24] It also has huge and complex facilities to reprocess plutonium from spent fuel and to create plutonium targets or "pits" for weapons. I was surprised when I read these facts, because I had naively thought that weapons labs only design the bombs, when in fact they are capable of making them.

There is now evidence to suggest that the DOE wishes to transfer mass production of bomb pits from the now closed and redundant Rocky Flats plant in Colorado to Los Alamos Labs.[25] Remember, the cold war is over.

Although no thorough environmental assessment of chemicals, or mapping of contamination, has been performed at Los Alamos, 1,857 sites around the lab are contaminated with

Table A-5 Summary of Hazardous Substances Released to the Environment at the Los Alamos National Laboratory

Contaminant	Air	Soil	Surface water	Ground water[a]	Sediment
Radionuclides		Americium-241		Cesium	Plutonium-239
		Beryllium-7[b]		Plutonium-238	Plutonium-240
		Cesium-134[b]		Plutonium-239	
		Cesium-137		Plutonium-240	
		Cobalt-57[b]		Tritium	
		Manganese-54[b]		Uranium	
		Mixed fission products			
		Plutonium-238			
		Plutonium-239			
		Plutonium-240			
		Sodium-22[b]			
		Strontium-90			
		Thorium[b]			
		Tritium			
		Uranium[b]			

Table A-5 Summary of Hazardous Substances Released to the Environment at the Los Alamos National Laboratory (Continued)

Contaminant	Air	Soil	Surface water	Ground water[a]	Sediment
Metals		Barium[b] Beryllium Cadmium[b] Chromium Copper[b] Lanthanum[b] Lead[b] Mercury[b] Nickel[b] Silver[b] Thallium[b] Cyanide[b]	Barium[b] Beryllium[b]	Barium[b] Beryllium	Barium[b] Beryllium
Inorganic compounds		Ferric chloride	Methylene chloride	Hexachlorobutadiene	
Volatile organic compounds (VOCs)		Hydrochloric acid Hydrofluoric acid Phosphoric acid		Methyl chloride Undefined VOCs	Acetone[b] Butyl acetates[b] Ethyl acetates[b]

Table A-5 Summary of Hazardous Substances Released to the Environment at the Los Alamos National Laboratory (Continued)

Contaminant	Air	Soil	Surface water	Ground water[a]	Sediment
		Sodium hydroxide			Methyl ethyl Ketone[b]
		Sodium thiosulfate			
		Sulfuric acid			
		Acetone[b]			
		Benzene[b]			
		Butyl acetates[b]			
		Ethanol[b]			
		Ethyl acetates[b]			
		Methyl ethyl ketone[b]			
		Tetrachloroethylene[b]			
Miscellaneous		Explosives[c]			Explosives[c]

[a]The groundwater medium at this facility essentially consists of perched ground water with no beneficial uses.
[b]The presence or potential contamination associated with this pollutant has not been fully determined.
[c]Examples of the explosives used at the site include Baratol, TNT, HMX, RDX, PETN, and Cytocol.

SOURCE: U.S. Department of Energy, Office of Environmental Audit, "Environmental Survey Preliminary Report—Los Alamos National Laboratory, Los Alamos, New Mexico," DOE/EH/OEV-12-P, January 1988; "Summary for the Los Alamos National Laboratory" submitted by U.S. DOE, Los Alamos Area Office, Environment, Safety and Health Branch on Aug. 3, 1990; and Thomas Buhl, Los Alamos Area Office, Environment, Safety and Health Branch, personal communication, Aug. 7, 1990.

plutonium, other radioactive isotopes, and hazardous chemicals. These areas include the county golf course, the town's main street, the Rio Grande River, most of the major canyons and mesa tops, the airport, and sixteen major dumps.

About 385 pounds of uranium and 88 pounds of plutonium were dumped in shafts created by test explosions. (Ten pounds of plutonium is fuel for one nuclear weapon). Radioactive emissions daily pollute the air, so breathing and eating locally grown vegetables and fruit would likely be hazardous to the community.[26]

More than three million curies of radiation have been released over the last ten years, an amount that is 250,000 times greater than the release from Three Mile Island.[27] Unmarked, top-secret trailer trucks routinely transport spent fuel from the lab's research reactor to unknown destinations. Plutonium accidents have happened, including a plutonium-laden truck that overturned in Colorado.[28]

Preliminary studies indicate that cancer rates in the local population have increased ten-fold in the last decade, including brain cancer and leukemia. Thyroid cancer is four times higher than the rest of the state, and Los Alamos County leads the state in childhood cancer and mortality.[29]

For decades, plutonium workers were provided with free cigarettes by the lab (smoking enhances the carcinogenic effect of plutonium in the lung). I suspect the lab did this to disguise the expected increase in lung cancer they knew would be caused by plutonium exposure. In 1990 alone, more than one thousand workers were exposed to radiation, and at least seven of these people inhaled or ingested plutonium. Preliminary studies indicate that this worker population has an increased incidence of bladder, brain, skin, kidney, thyroid, and uterine cancer and non-Hodgkins lymphoma.[30] But these figures may only be the tip of the iceberg because of the long latent period between exposure and the development of cancer, which can be anytime from five to sixty years; most of the radioactive material released into the environment at the weapons factories has occurred over the last few decades only.

Intimidation of workers is common when they raise safety concerns, yet the lab maintains that "the threat to public health is extremely slight or nonexistent."[31] No wonder I was not invited to speak at Los Alamos on the medical dangers of radiation in the early 1980s. But on the same visit to New Mexico, I was invited to speak at Sandia Labs close by, which was an interesting experience. I had a large audience of fascinated scientists[32], who thanked me for coming and pointed to their colleagues saying, "they needed to hear that."

Sandia National Laboratories established in 1945 and run by AT&T, designs and tests the non-nuclear components of nuclear bombs.[33] Radioactive elements, heavy metals, and organic and inorganic toxins have been identified in the surrounding air, ground water, and soil.[34]

The Feed Materials Production Center at Fernald, Ohio, in operation since 1953, confused the local population until just recently; because of its name, they thought the center made pet food. Instead, it produced uranium metal pellets for plutonium and tritium reactors at Hanford and Rocky Flats, and parts of weapons for manufacture at Rocky Flats and the Y12 plant at Oak Ridge, Tennessee.

Over the years, it has released, not inadvertently, 298,000 pounds of uranium dust into the air, 167,000 pounds into the Great Miami River, and 12.7 million pounds into earthern pits. Large silos on the reservation contain 9,700 tons or 1,600 to 4,600 curies of uranium, which is emitting high concentrations of carcinogenic radon gas. If the aging silo should collapse, huge quantities of radon gas would escape.[35]

Local people are developing highly malignant bone sarcomas and other cancers secondary to uranium deposits in their body organs.[36] Apart from the uranium, cesium 137, strontium 90, and alpha emitters contaminate the ground water; other toxic chemicals have been found in the soil, the air, and the surface and ground water.[37]

The Oak Ridge Reservation in Tennessee was the site for the first plutonium production reactor and for the early reprocessing of spent fuel. This facility includes the Oak Ridge Gaseous Diffusion plant (for uranium enrichment) and the Y12 plant, which produces lithium deuteride fusion components for nuclear weapons; it also converts lithium into tritium targets for the Savannah River reactors. The laboratory has totally inadequate documentation of the quantities and types of toxic releases over the years, and investigations of environment contamination are only just beginning.[38]

Since 1943, the plant has released thousands of pounds of uranium into the air, and radioactive and chemical wastes have been improperly stored or buried on site. There is extensive ground water, soil, and surface water contamination with isotopes such as plutonium, uranium, Americium 241, cobalt 60, cesium 137, radium 228, and others. In the ground water, these elements are combined with fourteen organic compounds that facilitate migratory patterns. Local reservoirs and streams that enter the Clinch River are severely polluted, and these water bodies service large off-site populations. The bottom sediment of the Clinch River is also contaminated, and the Watts-Bar Reservoir, a nearby recreational lake, is polluted with 175,000 tons of mercury as well as cesium.[39]

The workers are experiencing higher than normal levels of brain, kidney, lung, bone marrow, and lymph node cancer.[40] Epidemiology of the local population has yet to be done.

The Kansas City Plant in Missouri manufactures non-nuclear warhead components. Over time, it has released volatile organic compounds into the air and the soil. Surface and ground water are contaminated with a mixture of carcinogenic organic compounds, and the soil is contaminated with heavy metals and uranium. Mapping off-site contamination is on-going, and private drinking wells have yet to be assayed.[41]

The Pinellas Plant in Florida, near Clearwater, has the same function as the Kansas City Plant. Large numbers of toxic nonradioactive chemicals and tritium infect the air, soil, sur-face and ground water, and sediments. The deep Florida aquifer lies just beneath the plant, and although it is a major source of drinking water for a large population, the nature and extent of the contamination is not known.[42]

The Mound Plant near Dayton, Ohio, has, since 1948, made and tested nuclear weapons components; it also recovers and purifies tritium from old nuclear bombs.[43]

Mapping and assessing the extent of environmental pollu-tion may take eight to nine years, but already ground-water pollution has been found both on and off site. This is a very serious threat to human health, because this water is a sole-source aquifer for drinking water. Surface water and sediments are also polluted.[44] In 1969, a pipe carrying high-level waste at the Mound Plant broke, releasing plutonium into the Great Miami River and irrigation canals over a period of several years. The plant has also routinely emitted small quantities of plutonium and tritium into the air.[45]

Idaho National Engineering Laboratory was established in 1949 in an isolated location twenty-one miles west of Idaho Falls, to be the National Reactor Testing Station. Fifty-two reactors were built (the greatest concentration of reactors in the world). Reprocessing and nuclear waste storage also took place here. About thirteen reactors are still in operation. The laboratory also conducts safety research for civilian reactor programs, reactor development, nuclear waste, and spent fuel management research. It also is an integral component of the U.S. Navy's nuclear reactor program.

The processing plant recovers enriched uranium from naval reactors and from research reactors from other countries to be used as fuel for plutonium-producing reactors at the Savannah River. So the research reactor at Lucas Heights in Australia has been integral to weapons production.[46] If the Australian people knew this fact they would be very annoyed, because they have been reassured by their government that Australia does not participate in the production of nuclear weapons.

The ground water that connects with the Snake River aquifer is contaminated with iodine 129 (half-life of seventeen million years), plutonium 238 and 239, strontium 90, and tritium, as well as eight heavy metals, inorganic and organic carcinogenic compounds, asbestos, PCBs, sewerage, trichlorethalene, and fuel oil. This aquifer is the principal source of irrigation and drinking water for eastern Idaho. The soil is also heavily contaminated with the same chemicals.[47]

During the 1950s, over 350,000 curies of radioactive iodine and other fission products were released into the air.[48]

Rocky Flats, which was built in 1951, fifteen miles from the beautiful city of Denver, is an ecological catastrophe. It was designed to recover and recycle plutonium from old nuclear weapons and from plutonium-contaminated scrap, and to fabricate plutonium triggers for new bombs.

The recovery of plutonium involves chemical processing with toxic liquids, converting the recovered plutonium to gas, converting it to metal again, and then machining the metal ingots. Of course these activities produce vast quantities of contaminated waste, or scrap, including plutonium-infected liquids, old gloves, machinery, metal shavings, etc. For further details on plutonium processing, I refer you to the excellent document, *Plutonium Processing* (see note 12 of this chapter).

The plant also fabricated carcinogenic beryllium and uranium for weapons. It was closed in 1989, after a raid by seventy FBI agents, because gross safety violations were discovered. Most of the buildings are contaminated with plutonium, one grossly so, and a DOE study documented eighty-nine environmental and twenty management problems.[49]

Because plutonium is spontaneously flammable when exposed to the air, producing tiny particles of respirable size, and because of carelessness, there have been over seven hundred plutonium fires at Rocky Flats.[50] The major fire in 1957 burned the filters designed to trap plutonium. It burned for thirteen hours, and when the smokestack monitors were reconnected seven days after the fire, radioactive measurements were sixteen thousand times greater than "allowable" standards. No emergency action was taken to protect the public, and the fire released thirty to forty-four pounds of respirable plutonium into the atmosphere, which contaminated Denver and a major water source.[51]

Safety standards around the facility left much to be desired, and from 1959 to 1969, plutonium-infected machine oil leaked from three thousand storage drums, contaminating surrounding off-site areas. In 1973, tritium was accidentally released into the water supply of Broomfield—a small city of seventeen thousand people adjacent to Rocky Flats[52]

When the plant was built in 1951, the population of Denver was 567,000; it is now 1.4 million and growing fast. Most people live downwind from the plant.[53] Eighty thousand live within three miles of the factory, and houses are now almost adjacent to its back fence.[54]

The prevailing wind from the plant is toward Denver, and every gust blows over these people—a medical catastrophe beyond repair. Dr. Carl Johnson found the cancer incidence in Denver to be higher than normal, and even higher in suburbs near the plant.

In September 1982, sixty-two pounds of plutonium were discovered stuck in the plant's exhaust pipes and ventilation systems. As ten pounds is critical mass, a nuclear explosion could be triggered at any time.[55] In 1989, the Government Accounting Office deemed Rocky Flats the most serious envi-

ronmental threat at any weapons plant in the country, and the
land downwind experiences the highest concentration of plutonium in the country.[56] But real estate agents are still selling
houses and land in the contaminated areas.

The accompanying table documents the on- and off-site
contamination at Rocky Flats.[57] One hundred and eight inactive waste sites contain undocumented quantities and qualities
of isotopes.[58] Risk assessments to human health and the environment have never been done, however, even though the
DOE is charged with both the construction of nuclear weapons and the monitoring and prevention of resulting disease
processes.[59]

The incidence of bone marrow and lymphatic cancer, brain
and central nervous system cancer, and prostatic carcinoma is
increased amongst Rocky Flats workers.[60]

Although Secretary of Energy James Watkins said in January 1992 that "plutonium manufacturing operations at Rocky
Flats are now terminated" because the arms race was over,[61]
the DOE contends that "restart" is still a necessity both to
decontaminate the facilities and possibly to retrieve plutonium
from 308,000 pounds of contaminated material at the site.

Plutonium is so precious, like gold, the waste needs to be
mined! But 1,093 cubic yards of contaminated waste on-site is
not considered valuable enough to process.[62] What will happen
to this material and to the plutonium?

On April 5, 1993, the DOE announced that Rocky Flats will
never reopen for bomb production, and that $19.6 billion
would be shifted from nuclear weapons production to environmental "cleanup" and to the development of more efficient
energy sources.[63]

Hanford Reservation was established in dry, desert, farming
land beside the Columbia River in 1943 to create plutonium
for the Nagasaki bomb. Since that time, nine nuclear reactors
have been constructed, five reprocessing plants, and a plant to

Table A-11 Summary of Hazardous Substances Released to the Environment at the Rocky Flats Plant

Contaminant	Air	Soil	Surface water	Ground water	Sediment
Radionuclides	Beryllium Plutonium[b]	Americium-241 Gross alpha Gross beta Plutonium Tritium[a] Uranium	Plutonium	Cesium-137[a] Gross alpha Gross beta Strontium[a] Tritium[a] Uranium[a]	Cesium-137 Plutonium-239 Plutonium-240
Metals		Lithium[a]		Beryllium[a] Cadmium Chromium[a] Lead[a] Manganese[a] Molybdenum[a] Nickel[a] Selenium[a] Thallium	

Table A-11 Summary of Hazardous Substances Released to the Environment at the Rocky Flats Plant *(Continued)*

Contaminant	Air	Soil	Surface water	Ground water	Sediment
Miscellaneous	Laundry wastewater[c]	Disposed waste[d]	Disposed waste[d]	Disposed waste[d]	
		Friable asbestos	Friable asbestos	Friable asbestos	
		Oil sludge	PCBs[a,e]	Oil sludge	
		PCBs[a,e]		PCBs[a,e]	
		Sanitary sewage sludge[d]		Total dissolved solids	

[a]The present or potential contamination associated with current and past discharges of this pollutant has not been fully determined.
[b]Primarily due to past accidental releases.
[c]Significant releases of radionuclides into air may have occurred from 1969 to 1973 when radioactively contaminated sludges were dried at the facility's drying beds.
[d]There is a potential for soil, surface water, and groundwater contamination because current practices do not prevent low-level radioactive waste improperly disposed of in landfill designed for hazardous waste.
[e]PCBs = polychlorinated biphenyls.

SOURCE: U.S. Department of Energy, Office of Environmental Audit, "Environmental Survey Preliminary Report—Rocky Flats Plant, Golden, Colorado," DOE/EH/OEV-03-P, January 1988; "Federal Facility Agreement and Consent Order—Rocky Flats Plant"; and "Report on Federal Facility Land Disposal Review," October 1987.

convert plutonium nitrate to metal or plutonium oxide. The largest reprocessing plant, PUREX, extracted plutonium and uranium from spent fuel rods.

Hanford enshrines 60 percent by volume of the nation's dissolved high-level waste from reactor fuel, which is extremely radioactive. Because little or no regard was taken for "safe" disposal of this waste, corrosive prone, single-walled, one-million-gallon carbon steel tanks were sunk into the desert floor to receive the thermally hot, very radioactive, concentrated nitric acid solution from the PUREX plant.[64]

The corrosive factor of the nitric acid was understood at the time of construction, but all efforts were concentrated solely on nuclear bomb production; the waste would take care of itself—which it eventually did.

Of the one hundred and forty-nine tanks that hold 46 million gallons of the material, sixty-six have leaked. At least 750,000 gallons of liquid high-level waste have entered the soil, but this amount could be much larger. For instance, a 500,000-gallon leak from just one tank was not reported by Westinghouse or the DOE.[65]

The waste in some of these tanks is producing huge hydrogen bubbles, which could explode if ignited. This catastrophe would spread radiation far and wide, producing an American Kyshtym. Also, a chemical called ferrocyanide was added to some of the tanks in the past, and there is fear that this could combine with nitrate compounds at certain temperatures to produce an explosion with devastating consequences.[66]

Forty-eight newer, double-shelled tanks store another 11.4 million gallons of high-level waste. But apart from this well-documented waste storage, most of the waste was disposed of over the years with little or no documentation.[67]

Each 2.2 pounds of plutonium extracted at PUREX produced more than 340 gallons of highly radioactive liquid waste mixed with toxic waste, 55,000 gallons of less concentrated radioactive waste that was discharged to cribs in the soil, and 2.5 million gallons of contaminated water. Plutonium production at Hanford created one million gallons of high-level waste annually.

Hanford also discharged over two hundred billion gallons of low-concentrated radioactive water into evaporation ponds, burial pits, and seepage basins, all in close proximity to the Columbia River, which produces fish and recreation for the area.

Before 1970, plutonium-contaminated waste, packaged only in cardboard boxes and fifty-five-gallon drums, was dumped straight into the soil. Eleven of these particular waste sites contain 770 pounds of plutonium alone—enough to make seventy bombs.

Eight inactive reactors on the site, which need decommissioning, harbor ten thousand curies of activation and fission products, mainly carbon 14 and cobalt 60; while the PUREX plant, which closed in 1988 also needs decommissioning.

Some years ago, a trench, which contained 220 pounds of plutonium mixed with contaminated waste, had to be dug up at a cost of $2 million because Hanford managers were scared that it could reach critical mass and explode.

When scientists first decided to use the soil and trenches as waste-disposal facilities, they estimated that the radioisotopes would take 175 to 180 years to migrate to the Columbia River. Instead, the first contamination beyond the Hanford boundary appeared only eleven years later, and after PUREX commenced operation, a tritium plume made its way through the soil nine miles to the Columbia River in only nine years.

In October 1990, four thousand curies of tritium per year were being discharged to the river. The ground water beneath Hanford, which is contaminated with uranium, strontium 90, iodine 129, and plutonium, also communicates with the Columbia River.[68]

In fact, this beautiful river has been designated one of the most radioactive in the world, yet people still catch and eat fish without protection or warnings about health dangers. The quantities of isotopes consumed by local residents will never be known.[69]

As well as making its own waste Hanford is the repository of waste from other states and institutions. Six Polaris submarine reactors have been buried at Hanford, and authorities plan to bury one hundred more. In 1989 a 1,000-ton reactor vessel from Shippingport was buried on Hanford land, as were two hundred truckloads of nuclear waste from the same reactor.[70]

Irradiated fuel from Three Mile Island has been sent to Hanford and also to the Idaho National Engineering Lab, and the reactor internals have also probably gone to Hanford.

Irradiated fuel from the last remaining operating reactor at Hanford languishes in highly contaminated "swimming pools," which, in 1980, leaked fifteen million gallons of radioactive water contaminated with strontium and cesium.[71]

Hanford—America's nuclear cemetery—a vast dead satanic area, poisonous for millions of years.

And during the 1940s and 50s, 270,000 people were exposed to a radioactive cloud of half a million curies of iodine 131, which was purposefully released to the atmosphere by the authorities. This was called the "green run" by the authorities. Thirteen thousand children probably received doses of seventy rads or more to their thyroid glands.[72] But no official government health assessment or follow-up has ever been conducted upon this unfortunate population, let alone the provision of emotional or financial support.[73]

Instead the DOE has spent $39 million in legal fees opposing citizen claims for compensation and medical care for their radiation induced diseases.[74]

The incidences of lymphatic, liver, gall bladder, lung, and

bone marrow cancer are all elevated among Hanford workers.[75]

Savannah River Site was built by Du Pont in the 1950s and has been operated by Westinghouse since 1989.

I went to the Savannah River Site which is located in a high rainfall area amidst very poor farming communities, in the late 1970s. These people have unknowingly been growing and eating radioactive food for years, because large quantities of radioisotopes have escaped and settled on their land as a result of normal operations and accidental releases from five nuclear reactors and two large reprocessing plants. The site also includes a tritium facility and areas where plutonium, uranium, and neptunium are manufactured. Specifically, plutonium 238 is used to power space satellites.

All five of the reactors at Savannah River experienced numerous severe accidents, which were kept secret from the public for up to thirty-one years by Du Pont, the DOE private contractor. Some of the accidents were among the most severe ever experienced at a nuclear power plant.

There was a meltdown of a vital reactor component in 1970, which caused a huge release of radiation, and nine hundred men worked in a highly contaminated area for three months to clean it up. There was a leak in 1965 of 2,100 gallons of highly radioactive water, which almost precipitated a meltdown, and the core of another reactor almost melted because because technicians removed too many control rods—a carbon copy of Chernobyl. Exposure of the local population is not recorded. Despite the end of the cold war, the Department of Energy plans to replace Rocky Flats with Savannah River for the recovery of plutonium from old nuclear weapons; DOE also wants to build a new reprocessing plant.[76]

And until recently, as part of an expanded plutonium "mission," DOE had also planned to reopen one or two of the old Savannah River reactors to produce more plutonium and tritium and it requested money to process plutonium scrap from Los Alamos, Lawrence Livermore, and Brookhaven National Labs. Apparently during the 1980s it also wanted to incinerate plutonium contaminated scrap and to recover plutonium from the ash—surely a most hazardous procedure.[77]

The DOE has spent almost $4 million since 1988 to upgrade and reopen an old very unsafe reactor that lacks a containment vessel and an adequate filter system to make more tritium. But Secretary of Energy Watkins said in 1992, "We now need no actual tritium production until about the year 2005 to 2010."[78] Not only can tritium be recycled from old bombs, but President Yeltsin has called for reductions of strategic weapons on both sides from approximately 10,000 to 2,500. President Bush preferred 4,500, and the United States plans to dismantle two thousand bombs per year.[79]

In 1992, when Westinghouse, the new manager, attempted to restart a plutonium production reactor, tritium escaped into the river, contaminating local crops, and food-processing industries had to be closed downstream. In the past, similar accidents were usually ignored by the local community. It is totally inappropriate to build food-processing facilities downstream from this radioactive factory.[80]

I think the DOE anticipated that they could move their extremely dangerous operations from other facilities to the Savannah River Site because the surrounding population, unlike those in other areas of the country, has always been rather passive and relatively uneducated, possibly because the original inhabitants of the town were moved out lock, stock, and barrel, making it a company town. These conditions are now changing, as people realize the dreadful things that have occurred at the plant over the years. Like Hanford, Savannah River is a dreadful radioactive mess. There have been more than thirty serious reactor accidents that released large quantities of radiation over the years, some even involving meltdowns.[81]

In 1984, 1.7 million curies of tritium and krypton 85 were released routinely into the atmosphere, while millions more were released accidentally. Interestingly, superfund regulations now concentrate only on ground water contamination, while ignoring atmospheric releases, which are more dangerous.[82]

Savannah River stores an even greater quantity of high-level liquid radioactivity than Hanford, measured in curies; together they account for 90 percent of the nation's military waste, as measured by volume and radioactivity. Fifty-one subsurface, carbon-steel, high-level waste tanks hold thirty-five million gallons of high-level waste; some of these tanks have sprung leaks from rust holes,[83] while others, like the tanks at Hanford, are in danger of exploding.[84]

Apart from the high-level-waste fiasco, sixteen million cubic feet of low-level solid radioactive waste is buried in trenches at Savannah River and at 262 other waste sites scattered over the area. Some of the waste is even packaged in cardboard boxes, and hundreds of thousands of cubic feet of plutonium and actinide contaminated material are stored in interim facilities.

Scientists are very concerned that ground water contaminated with tritium, plutonium, and cesium may have communicated with the Tuscaloosa Aquifer, an important source of drinking water that is located beneath the states of Alabama, Georgia, and North and South Carolina; but the truth is that very little monitoring has yet been done to determine the extent, migration, and type of contamination.

We do know for a fact, however, that ground water at the complex is already contaminated with cesium 137, cobalt 60, plutonium 238 and 239, radium, ruthenium 166, strontium 90, tritium, uranium, and complex organic chemical solvents that enhance the radioactive migration.[85] Tritium has been found seventy miles downriver in Beaufort South Carolina.[86]

Preliminary epidemiological studies among the Savannah

River workers show a rise in cancers of the bladder, pancreas, and lymphatic systems.[87]

Pantex. Like the other bomb factories, this one is located in an isolated desert environment near a small town called Amarillo in Texas. During the antinuclear weapons protests in the 1980s, a lovely local catholic priest, Bishop Mathiason, spoke out forcefully against this satanic structure.

Built in 1942, its function was to assemble the component parts of the bomb that were sent from all the other bomb factories in the country. It also makes the chemical explosives designed to implode the plutonium trigger in the bomb, and it maintains and modifies the old bombs.

When the weapons are decommissioned, the plutonium pits will be stored in structures called igloos at Pantex.[88] A pit is a grapefruit-sized hollow ball of plutonium that, when compressed by conventional explosives, reaches critical mass and produces a nuclear explosion. There are sixty igloos at Pantex, each capable of holding 240 to 400 stacked pits. At the moment, 35 pits are added each week, so at the lower level one igloo is filled every seven weeks. Obviously, these igloos will be insufficient to hold all the pits from the dismantled weapons.

Pantex has thirteen assembly cells (gravel gerties), whose gravel ceilings are designed to collapse and contain the contents of an explosion. This facility can assemble or disassemble 1,500 to 4,500 warheads per year.[89]

Planned decreases in nuclear weapons will produce 72.5 metric tons of plutonium. Obviously, this material must be very carefully stored to protect it from theft, fire, or a critical mass accident.[90] Storage facilities will also need to be found for about 350 tons of highly enriched uranium obtained from decommissioned weapons. At the moment it is sent back to Oak Ridge, Tennessee.

By comparison, Russia will need to dispose of five hundred

metric tons of highly enriched uranium and sixty metric tons of weapons-grade plutonium. Options under discussion comprise burning the plutonium in a reactor to produce shorter-lived but highly toxic fission products, exploding the bombs underground, or storing the plutonium indefinitely.[91]

Russia sees plutonium as part of its national wealth and is loath to dispose of it permanently—if indeed this was possible. As they have inadequate storage facilities, they plan to build a $402 million storage space at the Tomsk-7 nuclear complex in Siberia, which will be partly financed by the United States.[92] Fancy America assisting the former Soviet Union to dismantle its nuclear weapons, an unthinkable situation five years ago. But dangers lurk below the veneer of good will. Russia offered in June 1992 to sell the United States tons of uranium from retired weapons; if the United States now, who else in the future?[93] Two thousand former Soviet scientists have an intimate knowledge of nuclear weapons design, and 3,000 to 5,000 have worked in uranium enrichment or plutonium production.[94] Most of these people are now unemployed and very poor. Will they offer their services to Iran, Iraq, etc.?

American options under discussion include fissioning the plutonium in commercial reactors, using it in breeders to make more, storing it above ground, or burying it.[95]

None of these options remove the threat of radioactive food, air, and water for the rest of time.

Surprisingly, there is no official verification method under the INF and START treaties to make sure that the United States and Russia are actually disarming their weapons. In fact none of the disarmament treaties call for dismantling the bombs, only for destruction of their delivery systems. A U.S. START negotiator recently stated that he did not think that destruction of warheads would ever be included in disarmament treaties, because then the United States would have to reveal the exact details of their warheads, and they would never do that.[96]

No environmental exposure or risk assessments have been conducted at the Pantex site, and ground-water examination is inadequate. Surface water and sediment are believed to be contaminated, and alpha and beta emitters have been detected in the soil.[97] Like all the other nuclear weapons factories much monitoring remains to be done, because almost none was done in the past. Elevated rates of leukemia, lymphosarcoma, and brain tumors have been found in Pantex workers.[98]

It is important to note that workers and scientists employed by the weapons facilities who raise significant safety issues with the press are often harassed by the DOE subcontractors. For instance, when Dr. Greg Wilkinson, an epidemiologist at Los Alamos National Labs, circulated a draft paper showing an excessive incidence of brain cancer amongst Rocky Flats workers, one of his supervisors told him that his findings would "shut down the nuclear industry"; another said he should do research "to please the DOE, your sponsors, not satisfy peer reviewers." He was pressured to withdraw his paper from publication in the *American Journal of Epidemiology,* and after it was published, the epidemiology group at Los Alamos was downgraded in status.

Another scientist resigned from the Oak Ridge National Laboratory after experiencing intimidation when he discovered extremely high levels of mercury in vegetation in a creek that passed through Oak Ridge National Labs. Subsequent study found that 2.4 million pounds of mercury had been dumped in the creek, together with significant quantities of arsenic, PCB, and organic solvents.

Often scientific data obtained by federal government scientists is subject to delays, censorship, intimidation, and interference with the constitutional rights of the scientists.[99]

No wonder there have been no adequate studies on the human health consequences of bomb making. I would venture to suggest that this is the biggest cover up in human history.

We must also remember that in this chapter, I have not discussed the waste accruing at over one hundred civilian nuclear power plants across the United States. The radioactivity of military high-level waste is only one-sixteenth of that contained in commercial spent fuel rods.[100]

Nor have I discussed the health problems of the former Soviet weapons makers. Nine hundred thousand people have worked in the industry,[101] and they, like their U.S. colleagues, were poorly protected against the medical hazards of radiation exposure. Mr. Minayer, who worked on a warhead assembly line for fourteen years said that he and his fellow workers wore rubber gloves and wiped their hands clean with alcohol. After the shift he said, "we returned to the dormitory, slept a bit, swept the fallen hair from our pillows, and went back to the workshops." Those who complained about the dangers of radiation were told that military men had to endure their fate in silence.[102]

Chapter 11

Waste Cleanup

In the 1950s, the United States possessed only a few hundred curies of radioactive waste. By 1984, it had accumulated 14.7 billion curies stored in interim centers; by the year 2000, experts predict a total of 42 billion curies—overkill for every human on the planet.

The cost of maintaining a permanent watch on old, disused, radioactive nuclear power plants and their waste has been transferred from the utilities to the taxpayer. These costs have never been factored into the cost of nuclear power.[1]

In fact, there are absolutely no viable solutions to the safe storage of radioactive waste, and as a scientist I predict there never will be. We are leaving a terrifying legacy to future generations—just, so the argument goes, that we could live with electricity, when for three million years our ancestors survived without it. The truth is that alternative power sources are plentiful and cheap, but have not been exploited or mass produced.

For the last fifty years, short-term, irresponsible waste-disposal decisions were made by the DOE, the NRC, the White House, and Congress; but future generations who will inherit our radioactive legacy were never considered.

PRESENT DOE WASTE-DISPOSAL PLANS

It has been estimated that cleanup of radioactive and mixed wastes will cost more than $100 to $150 billion and will take thirty years to complete.[2] This estimate is fallacious because the extent of the contamination is unknown, technologies for "cleanup" do not exist, and there are very few qualified people to conduct the testing. Therefore, no data exists to estimate the cost or duration of the cleanup.[3]

Most of the waste cleanup at this time has been "privatized" or subcontracted to corporations by the DOE. In fact, Westinghouse, which runs several of the weapons factories, is involved at various stages of the radioactive waste saga, including high-level waste disposal at Hanford, Idaho; Savannah River; and West Valley—the latter facility is a failed commercial reprocessing site.

So Westinghouse, which makes the bombs and the waste, then obtains lucrative government contracts to "clean up" the waste. As we say in Australia: "nice work if you can get it."

The nuclear industry figures that it cannot build more nuclear power plants, and the DOE cannot modernize its nuclear weapons production facilities until the public is reassured that safe disposal of radioactive waste is guaranteed.

So desperate and manipulative efforts are underway to coerce and pressure states, Native American tribes, and the rest of the public to accept and store deadly cargoes of radioactive material. The waste proposal plans lack scientific credibility on

close inspection and many are immoral. Descriptions of these
plans follow.

BELOW REGULATORY CONCERN

As the nuclear industry ages, with decommissioning of major
old nuclear reactors imminent and no solution to the waste
problem in sight, the Nuclear Regulatory Commission (NRC)
decided in 1986 to try a new approach to limiting the waste
problem: declare a large percentage of the total volume of
radioactive waste to be legally nonradioactive, or Below
Regulatory Concern (BRC). This linguistic detoxification, if
implemented, would have potent radioactive materials from
nuclear reactors containing isotopes, such as strontium 90
(half-life 28 years), plutonium (half-life 24,400 years), cesium
(half-life 30 years), nickel 59 (half-life 80,000 years), and iodine
129 (half-life 17 million years), dumped into sewers, landfills,
aquifers, and garbage dumps. In fact some so-called low-level
waste will obviously be as hazardous just as long as high-level
waste, which is classified only as irradiated fuel. And to make
matters worse, chelating agents are used as decontaminating
chemicals to clean out reactor pipes by dissolving insoluble
isotopes; these of course are buried in landfills with the deadly
isotopes described above. The practice of burying combined
chelating agents with radioactive waste speeds up the migration process through the soil.⁴

The BRC waste can also legally be incinerated, a process
that will contaminate the air with radioactivity. Obviously,
these uncontrolled random methods of disposal will also contaminate the air, soil, and water supply.

No labeling, warning, notification, or disclosure will be required, and no records will be kept. Waste from nuclear power

plants such as contaminated gloves, pipes, sludge, resins, oil, concrete, and large nuclear power plant components will become common garbage, and mixed radioactive and chemical waste will be disguised as hazardous, not radioactive waste.

The DOE is bound to join in and send large amounts of unmonitored military radioactive waste to common public facilities for disposal at no cost and little inconvenience. This is an aberration of current practice, which designates that all waste must be strictly monitored, recorded, and supervised, making present waste disposal very time-consuming and expensive.

As nuclear reactors age, nuclear waste will markedly increase, and reactor waste will comprise up to 99 percent of the radioactivity in low-level waste. The entire radioactive waste program is in a state of crisis, and commercial radioactive waste generators such as power reactors are quickly running out of places to send their waste.

The nuclear industry says that BRC could save about $82 million per year.[5] Under BRC, the NRC identifies "waste streams" emanating from radioactive facilities; in 1988 the NRC decided that it was acceptable for each person to accrue up 10 millirems per year from each deregulated waste stream, but that in certain situations up to 100 millirems per year would be acceptable. We must remember that 100 to 170 millirems is the amount of radiation already accrued per year from background radiation exposure. So the NRC would have us double our exposure, and their estimate for this dose is a "hypothetical lifetime risk of fatal cancer at one in 285 people for an annual exposure rate of 100 millirems."

All these arbitrary calculations for waste streams will be recorded on distant computers but there will be no actual "hands on" monitoring or regulation of what waste streams go where. And there will be no limit on the number of waste streams emanating from or going to a particular facility. Many "streams" could therefore be deposited in a single garbage dump, causing a buildup of concentrated radioactive waste, which will, over time, expose workers, drivers, children, and the general public to dangerous doses of radiation.[6]

Local politicians have been alerted to this danger, and fifteen states have already passed anti-BRC regulations, but their stand may be invalidated by a recent Supreme Court decision that grants civil rights to corporations. This new law gives corporations the right to sue a state and, if they win, to demand reparation. Tragically, because most corporations run entire departments of lawyers, many states because of their tight financial situations are already backing off from attempting to enforce their own laws.[7]

Under the new BRC policy statements put out by the NRC governing radioactive waste disposal, health compensation is not covered and, further, localities and states will have to bear the financial responsibility for eventual decontamination.[8]

This BRC policy could become a momentous, uncontrolled environmental catastrophe and an unmitigated public health disaster, at a time when radiation experts are calling for a reduction in allowable radiation exposure to the public and radiation workers of between four and thirty times because of newly analyzed data from the atomic bomb cohort.[9]

Under BRC, the NRC also proposes to recycle radioactive waste into consumer goods such as frying pans, jewelry, belt buckles, home appliances, bathtubs, furniture, kitchen sinks, surgical prosthetic implants, copper wire and piping, copper and nickel coins, and iron tonic.[10] The NRC in its magnanimity did exclude radioactive toys, because they recognized that children are forty times more radiosensitive than adults. They said that toys have "little or no benefit to society."

Radioactive topazes and amethysts are currently on sale to the public, as are radioactive false teeth, watches, fire alarms, antistatic devices, some ceramics and glassware, bell pushers,

and exit signs. All of these devices are already in wide circula-
tion. They contain a variety of isotopes including Americium
241, plutonium 238, radium 226, tritium, promethium 147,
krypton 85, cesium 137, uranium, and thorium.[11] Obviously
the nuclear industry already makes a tidy profit by selling its
dangerous waste.

FOOD IRRADIATION

Cesium 137 is the isotope used to irradiate food. This practice
is dangerous because the canisters that contain the cesium
expose workers and drivers to high-level gamma radiation. Although the irradiated food is not radioactive per se, it does
contain "radiolytic" chemicals, produced when gamma radiation interacts with food—none of these materials have been
tested for carcinogenicity. But because the seal sometimes fails
in the racks of cesium, the water soluble cesium 137 contaminates the water that is used to shield the device. The shielding
water then becomes quite radioactive, and when the device is
raised and lowered, splashing occurs that actually contaminates
the food.[12]

Two hundred and twenty-seven cesium 137 capsules, and
forty-three strontium 90 capsules were dispatched over the
years by the nuclear industry for commercial use in food
irradiation and for the sterilization of medical supplies. The
NRC in fact gave final approval to large irradiators in April
1993, and the FDA has approved irradiation of poultry, beef,
fruits, vegetables, spices, etc. A very dangerous move, and one
about which the public must be adequately informed.[13]

Distribution of high-level waste to an ignorant public can
pose dreadful problems. In 1987, in Goiania, Brazil, a family
found a canister containing 1,400 curies of cesium 137 at a

rubbish tip, which had been dumped by the local hospital. They were fascinated by this magical material, which glowed blue in the dark. They ate it and smeared it on their faces, and within days they all died of acute radiation sickness. The whole town was contaminated, never to be completely clean again, and many people were irradiated.[14] This is a frightening scenario for the future, if present practices are allowed to continue.

But other dreadful proposals are on the drawing board or are already in practice in the United States.

The NRC proposes to allow public access to radioactive buildings and land. It also wishes to allow the disposal of radioactive material into water.[15] And it has encouraged recycling of "slightly" radioactive scrap metal for steel production. Waste containing small amounts (<0.05 microcuries) of tritium and carbon 14 is already legally dumped into drains and municipal landfills. In Oklahoma, the NRC permits uranium and radioactive, nitrogen-rich fertilizer from uranium enrichment to be used in pastures for cattle.[16]

In 1982, the NRC redefined allowable radiation levels in transuranic waste to be over 100 nanocuries instead of 10 nanocuries. This new rule then permitted 38 percent of plutonium-contaminated waste to be reclassified as low-level waste. And now that BRC is in operation, "low-level" plutonium-contaminated material can be dumped with impunity into the public ecological domain.[17]

But in reality, unofficial BRC policies were already in effect for five years before the 1985 BRC legislation. The DOE has probably illegally sent military radioactive waste to twenty-six landfills and incinerators across the country since 1980. DOE private contractors, like Martin Marietta, may unilaterally have sent massive amounts of nuclear waste for unregulated disposal. If they did, who else?

In 1991, Savannah River nuclear waste was burned in the

Rollins hazardous waste incinerator near Baton Rouge. Some of this material contained technetium 99—a mobile isotope with a half-life of 212,000 years. But the DOE refuses to release documented radiation levels of the waste. Therefore it was almost certainly more radioactive than they care to admit. DOE says it stopped this practice in April 1990, but does it know what its contractors are doing now?[18] This is a time when the cancer merchants should be jailed for white-collar crime. This plan might have remained an obscure policy and even been implemented were it not for two women: Dr. Judith Johnsrud and Diane D'Arrigo, who were playing watchdog to the NRC at this time (1986). These two took the message of the BRC concept and all of the destruction it could bring to the grassroots, antinuclear and environmental movement, to the media, to state legislators, and to the Congress.

As a result, Congress passed a bill in August 1992 containing a provision (Title 29, Subpart A) that revoked both the 1986 (deregulation of waste) BRC policy and the 1990 Expanded Policy (deregulation of waste, materials, practices, treatment, consumer, and construction products, i.e., any radiation that complied with dose threshold limits defined by the policy). And it affirmed states' authority to set stricter-than-federal standards on the basis of radiological hazard for any radioactive waste exemplified by the NRC after the date of the act.[19]

However, not to be outdone, the NRC has since proposed an alternative to BRC—Enhanced Rulemaking on Residual Radioactivity (ERORR), or "son of BRC." Their goal is to set a level of radioactivity that the nuclear industry is allowed to leave behind when a contaminated site or decommissioned site is opened for "unrestricted use," i.e., daycare centers, vegetable gardens, and so on. Clearly the NRC has done an end run around BRC and will introduce more frightening scenarios in the future.[20]

CENTRALIZED DRY STORAGE OF IRRADIATED FUEL

In 1987, Congress passed a law requiring the DOE to construct a network of concrete bunkers and buildings temporarily to "dry store" tons of extremely radioactive, spent-fuel rods from civilian reactors. Land owned by Native American tribes in deserted states has been targeted. Some tribes have accepted $100,000 as "free money from the Feds" since this first grant had no strings attached, to allow DOE to conduct initial investigations. More money was promised if DOE proceeds with the plan. This policy is being pursued by one David Leroy, U.S. Waste Nuclear Negotiator, for a program called "Monitored Retrievable Storage Facilities" (MRS). At present, there are no nuclear reactors on Native American land.

Leroy gave a speech to the National Congress of American Indians in December 1991, in which he quoted extensively from the beautiful, poetic, environmental speech partly attributed to Chief Seattle. I consider that the use of this speech to justify the MRS enterprise is sacrilegious. Apparently some Native American tribes are in favor of the proposal and some are against. Usually the tribes who support the proposal are those where a single individual has held brutal control of the Tribal Council so that no one can effectively challenge him. In these situations it is known that nuclear industry interests have been in there for a long time, and much of the DOE grant money actually goes to "industry consultants."[21] Governor Mike Sullivan of Wyoming wisely refused further permission to base the high-level radioactive material in his state after the initial investigations were completed.[22]

The whole point of this dilemma, "though this be madness yet there's method in it" is that no democratically elected body in state government would accept such a dangerous facility, but Leroy says that governors have no authority over what a tribe does, even if their lands are surrounded by the state. To date, governors of eight other states have gone on record to oppose MRS in their state.[23]

OFFICIAL HIGH-LEVEL WASTE AND TRANSURANIC WASTE DISPOSAL

The DOE proposes to deposit some of its military transuranic waste in excavated salt caves below the desert floor in New Mexico, and to dispose of its high-level waste by encasing it in glass logs and putting it into man-made caves in Yucca mountain, Nevada. Vast problems accompany each proposal.

TRANSURANIC WASTE

Until 1978, transuranic waste (TRW) was buried at random in vast quantities beside the weapons factories, but documentation was sparse or nonexistent. After 1978, adequate documentation of TRW became available.[24] So much of this extremely dangerous plutonium-contaminated material is lost and therefore irretrievable. Of course, the transuranics, plutonium, neptunium, curium, Americium, etc, are all alpha emitters, and most have extremely long half-lives.

The TRW waste load includes metal, equipment, rags, gloves, paper, clothes, dirt, gravel, sludge, etc. Sixty percent of it is mixed with hazardous chemicals such as organic solvents, which facilitate the environmental migration of plutonium and its dangerous alpha-emitting relatives. But, the Environmental Protection Agency has granted an exemption to the DOE so that it does not need to treat this hazardous chemical TRW waste in the prescribed manner because it is so dangerous. From my reading, the powerful DOE often overrules the EPA.[25]

Exotic waste disposal plans include incorporating transuranic waste into glass, or calcining it (converting it to a soluble salt).[26] It is not a good idea to incorporate plutonium into a soluble material, because sooner or later it will leak and escape from its confines when exposed to water. But in reality, each of the above proposals will expose large numbers of workers to plutonium during the production processes, and the end products would still require disposal. Some people even want to incinerate the waste, which would broadcast plutonium far and wide.

Given the options, it may be in fact safer to transport the TRW intact from the weapons facilities and present storage sites. This last option means trucking fifty thousand drums containing plutonium and its dangerous radioactive relatives every year across the vast network of U.S. highways, where they will be vulnerable to accidents, fires, human frailties, etc. Adequate disposal time would take twenty-five years, and the DOE has yet to construct a cask that meets its own safety requirements.

Some of these drums emit such high-level gamma radiation that they can be handled only by remote control. Even so, all drivers and handlers will be exposed to some carcinogenic radiation.[27] The trucks will pass through twenty-three states, forty major population centers, and twenty-six Indian reservations.

DOE plans to bury six million cubic feet, or one million barrels, of plutonium waste at the Waste Isolation Pilot Plant (WIPP), in New Mexico, in caverns excavated in a huge salt deposit 2,150 feet underground.[28] This particular project was initiated after Congress explicitly prohibited this state from exercising any veto authority or control.

The National Academy of Sciences initially suggested storing radioactive waste in salt beds in 1959, because intact salt deposits indicate a geological stability of millions of years. However, a pressurized brine reservoir has been discovered beneath the WIPP site, and the Rustler Aquifer, which drains into the Pecos River, supplying water to residents in Texas and New Mexico, lies above excavated caverns. Of course, man-made holes in the salt bed during the WIPP construction will readily permit penetration of water; once water enters the salt bed it will quickly dissolve the salt, causing erosion of the drums and subsequent release of the plutonium and its relatives. So in fact these carefully laid plans are in vain.

Recently, it was discovered that gas will accumulate in the decaying waste because it is mixed with volatile organic solvents. This reaction will induce high pressures, which could not only rupture the drums but also fracture the surrounding rock and salt, facilitating radioactive escape. Located nearby are significant oil and gas reservoirs. If anyone drilled into these natural formations years hence, they could hit a pressurized radioactive brine reservoir and create a plutonium geyser.

Interestingly, WIPP is planned to accommodate only 20 percent of America's TRW. Seventy percent of the space is to be reserved for the waste generated over the next twenty years during nuclear weapons production.[29] These plans are obviously archaic even before they are begun.

HIGH-LEVEL WASTE DISPOSAL

Yucca Mountain in Nevada was chosen by Congress in 1987 as a repository for the nation's high-level waste. Both commercial nuclear power plant waste and millions of gallons of military reprocessed waste will be included.[30] Already one billion dollars have been spent on this project, and twenty-five billion more have yet to be spent excavating caves in the mountain. But the facility will not be ready to receive this material until the year 2010. In the meantime, DOE hopes that Monitored Retrievable Storage will be in service.[31] If the DOE can solve the problem of high-level waste storage, they believe they can then proceed with the construction of their advanced, "intrinsically safe," passive nuclear reactors.

There is such strong opposition in Nevada to this radioactive mountain that the American Nuclear Energy Council began an $8.7 million campaign in December 1991, which they called the Nevada Initiative, to convince the citizens that nuclear waste is safe, it won't leak, it can't spill or explode, there is no possibility of transportation accidents, that the repository is inevitable, and that Nevada will lose other federal government funding if it opposes the project. Despite the onslaught of this propaganda, 80 percent of the population still oppose the site. Meanwhile, the DOE claims that they are "studying" the site, but really they are continuing to prepare it for their nefarious plan.[32]

Not to be deterred, the nuclear industry plans to deploy "Scientific Truth Response" and "Media Response" teams as truth squads to respond to "scientifically inaccurate, misleading, or untrue allegations" that are published or aired about nuclear issues or Yucca Mountain.[33]

But to foil their plans, in June 1992, a 5.2 Richter scale earthquake caused $1 million damage to the DOE building just six miles from Yucca Mountain, two days after a 7.4 magnitude quake in Yucca Valley, Southern California.[34]

DOE has said that the site must be disqualified if there is renewed volcanic activity, if seepage of waste could pollute underground water, or if intrusion by future generations is likely.[35] At least thirty-two active faults pass within twenty miles of the mountain, and one, appropriately called the Ghost Dance, transects the proposed site.

The rock of Yucca Mountain is fractured, and there are a myriad of tiny cracks running throughout the stone. The gas carbon 14, with a half-life of 5,730 years, will be released during the decay process of the high-level waste. When it was discovered by scientists that Yucca Mountain could not retain radioactive gases, and so would not meet the EPA's high-level waste standards, Congress decided to exempt the facility from these standards. Further, in a closed-door session without any previous debate, they directed EPA to write standards that the mountain could meet, to be based on a report that a nonregulatory body, the National Academy of Sciences, was directed to write.

The National Energy Policy Act directs the National Academy to look only at the impact of the facility on any single individual, rather than looking at the global impact over the hazardous life of the waste that will be placed there. Carbon 14, released by the fractured stone of Yucca Mountain into the atmosphere, will contribute to global fall out, with 70,000 curies as the projected release.[36]

The dump is planned to hold seventy thousand metric tons of potent, high-level waste, and a large release, according to some experts, could cause environmental damage on the scale of nuclear war. Ground water beneath the mountain could rise as it has in previous geological time frames, flooding the hot canisters and blowing the top off the mountain. Alternatively, contaminated water could seep into the ground water and spread to Death Valley, producing radioactive hot springs. U.S. Geological Survey scientists compare the Yucca Mountain

Program to "NASA before the Challenger."[37]

One spent-fuel assembly contains ten times more long-lived isotopes than the Hiroshima bomb, and 140,000 of these assemblies will be transported across the United States to the mountain. Temperature inside the repository will be above boiling point for 1,250 years, and the temperature inside the bore hole of the rock will be 527 degrees Fahrenheit and in the canister, 662 degrees Fahrenheit. Such heat could induce rock fractures and fault movement.[38]

Transport of seventy thousand metric tons of civilian, spent-fuel rods over thirty years is predicted to involve fifty accidents per year, three being severe with possible radioactive release. At this time, there is no prohibition of these shipments through highly populated areas nor during severe weather conditions, and human error has not been factored into the safety calculations. All of the eleven casks currently used by the DOE for high-level waste transportation have been found to be defective.[39]

DOE has already established that above-ground dry storage of spent-fuel rods at reactor sites is safe for one hundred years, provided they have been cooled in the swimming pools for at least ten years.[40] This known fact gives scientists the latitude to adequately decide how to store radioactive waste safely for several million years (if this is possible) and to choose more suitable sites than those presently proposed.

But it also makes the public very nervous to live near the radiation equivalent of several thousand Hiroshima-sized bombs. Out of site is out of mind. That's why DOE wants to put these fuel rods in Yucca Mountain quickly, hoping the public will then relax and condone an ongoing nuclear power industry. An exercise in sheer futility when one considers the costs, dangers, and CO_2 production of nuclear power and the potential cheap alternative energy techniques just waiting for mass production.

Not only will Yucca Mountain accommodate spent-fuel

rods but DOE plans to deposit the high-level military waste at this site also. And because the liquid waste tanks at Hanford and Savannah River threaten surrounding communities with Kyshtym type explosions, some urgency exists to expedite the problem.

Present plans call for decanting the supernatant liquid from the top portion of the tanks, and removing the intensely radioactive cesium 137 and other isotopes from this liquid by precipitate hydrolysis, then mixing the remaining fluid with concrete to make a solid known as "grout." This process actually increases the waste volume, but it is more easily handled as a solid than a liquid. The grout will be stored temporarily on site at Hanford, Savannah River, and West Valley.[41] The grout contains, among other isotopes too numerous to mention, technetium 99, with a half-life of 210,000 years. Grout is not a stable substance and will leak isotopes over the years. Plans for permanent disposal are not clear.

The remaining intensely radioactive waste will be heated in a ceramic container with ground borosilicate glass to produce a vitrified glass log, weighing 3,700 pounds, containing ninety-four pounds of radioisotopes, emitting 234,000 curies of radiation and it will be intensely hot.[42] Placed in a metal canister with three-eighths of an inch thick walls, the expected lifetime of this device will be one hundred to one thousand years. But in fact, the integrity of the canister must be maintained for 500,000 years for plutonium and very much longer for iodine 129, which has a half-life of sixteen million years. In this context, the Swedes use a thicker-walled canister, of ten centimeters, which is considered to be safer although obviously still not adequate.[43]

Most European countries plan to cool their glassified waste on site for fifty years to allow some decay of the radioactive isotopes before they consider permanent storage.[44] The very high-energy alpha emitters in glass will produce microscopic cracks called devitrification, thereby causing embrittlement, corrosion, and leaking of the isotopes from the glass.[45] Hence, some scientists have grave concerns about the long-term "performance" of the borosilicate glass logs and their containers that will be used at Yucca Mountain.

Glassification will be done at four locations, Savannah River, Hanford, Idaho National Engineering Laboratory, and West Valley, New York, under the sponsorship of Westing house.[46]

The problem is that by processing, glassifying, and moving these wastes around the country, thousands, if not millions, of individuals will be exposed to cancer-causing radiation. It's like stirring up a hornet's nest. Would it not be better to let sleeping dogs lie, leave the waste where it is presently situated, evacuate the surrounding populations from Denver, Richland, Amarillo, etc., and remove the risk of ongoing radiation exposure to millions more people?

It is important to know that the planned "cleanup program," although supervised by the DOE, will be almost totally conducted by private contractors who will design, engineer, and perform the environmental projects. If the weapons industry can be used as a typical example, contractors routinely overestimate cost. So the estimates of cleanup are grossly inflated. The same companies that rorted the American public to build the weapons are now rorting them again to "clean up" the waste those companies made.[47]

The "cleanup" plan rivals preparations for the first moon landing, according to Matthew Wald from the *New York Times.* The first five-year plan includes projects in thirty-four states and Puerto Rico.[48] I remember feeling a vague sense of discomfort when I encountered two sweet young women from Puerto Rico in an elevator of a U.S. hotel who were attending a DOE conference. I knew something fishy was afoot. Will some of the U.S. nuclear waste be stored in Puerto Rico?

LOW-LEVEL WASTE DISPOSAL (WHICH IS NOT REALLY
LOW-LEVEL AT ALL, AS PREVIOUSLY EXPLAINED)

In 1980, the Congress passed the Low-Level Radioactive Waste Policy, mandating that by 1993 every state must accept liability and responsibility for all low-level waste produced by the nuclear industry in that state, letting the utilities off scot-free. In 1988, California signed an agreement with Arizona and North and South Dakota to accept their low-level waste and store it in California, and the state government strangely decided to allocate responsibility for the dump to the Department of Health Services (DHS). In 1985, DHS chose U.S. Ecology, a subsidiary of Browning Ferris Industries, a notorious private firm, to build the dump. (Radioactive waste burial dumps operated by this firm in Sheffield, Illinois; Beatty, Nevada and Maxey Flats, Kentucky, are both leaking.) U.S. Ecology then paid the League of Women Voters to promote the site by holding hearings across the state. The league was also paid $274,000 in Nevada for a similar campaign.[49]

The selected site in California is eighteen miles west of Needles and thirteen miles west of the Colorado River, which is the water source for millions of people living in the Los Angeles basin, San Diego, San Bernadino, Phoenix, Tucson, and for food growers in the Imperial Valley. It is located six hundred feet above a pristine underground lake the size of Lake Tahoe. Eighty-five percent of the waste, which will be dumped in unlined earthern trenches with absolutely no barrier to migration of the isotopes, will emanate from nuclear reactors and scientific research with only 0.5 percent coming from medical wastes.[50] Maybe the carcasses of one thousand frozen beagle dogs contaminated with radium and strontium 90, which have been stored at the University of California for thirty years will also be dumped at Needles.[51]

The desert environment is subject to fierce winds, which will carry the radioactivity to distant communities, and to flash flooding. When rain falls on desert, the water migrates deeply very quickly carrying with it the toxic radioisotopes; and should it evaporate in the heat, soluble isotopes evaporate with it.

In this chapter I have only skimmed the surface of the radioactive waste problem. I invite you to read the references, some of which have been prepared by voluntary concerned citizen groups who are performing an excellent service by researching and publishing heretofore unobtainable data. More is yet to be revealed.

To summarize the present status of "low-level" waste disposal: current federal law requires that every state either open a radioactive "low-level" waste dump, or form a compact with a host state to house the dump. The deadline set was January 1, 1993; as of February 1994 no dumps or compacts were in place. But the fourteen proposed new commercial waste dumps have been held at bay by local communities.

The compacts and their hosts are:

Northwest: Washington (Richland existing dump)
Rocky Mountain—contracts with Northwest
Southwest: California
Central States: Nebraska
Southeast: currently using Barnwell, South Carolina, but have selected North Carolina for the new host for the proposed dump
Central Midwest: Illinois
Midwest: Ohio
Appalachian: Pennsylvania
Northeast: both Connecticut and New Jersey

All of these compact dumps, except those that will continue to use Richland, Washington, are proposed facilities only. None of them have successfully met the federal milestones, but you may be sure that almost all of them will be located in very poor areas.

The other states that have not yet joined a compact and are either negotiating to do so now, and/or are planning their own facility are Maine, New Hampshire, Vermont, New York, Massachusetts, Rhode Island, Washington, D.C., Michigan, and Texas.

Maine and Vermont are negotiating for compact affiliation with Texas, with Texas as the host dump state. Connecticut is offering $100 million to get into the deal.

Nationally 88.5 percent of the radioactivity and the majority of the volume of commercial "low-level" waste comes from nuclear reactors. When the decommissioning of the aging reactors is factored in, it will be 99 percent of the radioactivity and a very large majority of the volume.

Academic research and medical practice produce less than 2 percent of the radioactivity in "low-level" waste, yet a small, special interest group in the medical community is backing the push for fourteen new dumps. They claim that hospitals will have to shut down and that cancer treatments will stop. The truth is that medical diagnosis and cancer treatment result in very short-lived waste that decays in a matter of hours or days to undetectable levels.

These doctors also form a lobbying group for the continuation of the nuclear power industry, claiming that nuclear power produces the isotopes used in medicine. All medical isotopes can be made by a cyclotron, which produces no radioactive waste, and many presently are. Further, we should not be using isotopes in medicine that require a nuclear reactor for production. This is aberrant logic, because for the very few cancer and other patients who are helped by these isotopes, thousands or millions more will die in the future from the radioactive wastes made in their production.[52]

SOME SUGGESTED SOLUTIONS TO THE RADIOACTIVE WASTE PROBLEM

1. Close down all nuclear reactors immediately—both civilian and military.

2. Allocate sufficient funds on a priority basis to correctly identify the quantity, composition, and extent of the contamination. The DOE has only just begun to address this problem after all these years, and it will be more than five years before decent initial data are obtained at all sites.[53]

3. Notify the populations living around each radioactive and chemical cauldron about the present and past dangers. Explain the type of toxins involved, the routes of contamination—air, water, and bioconcentration in the food chain. Involve them in the decision-making and enlist their help, advice, and expertise.

4. Establish as a priority, scientific epidemiological studies on the seven hundred thousand or more nuclear workers and on all potentially exposed populations. This project will involve tracking individuals from the estimated time of exposure to death and clearly defining the accurate cause of death.

5. All DOE records of radiation exposures and releases and all data from private contractors must now be opened to public scrutiny.

6. All responsibiliy for health, epidemiology, and cleanup must be immediately removed from DOE jurisdiction. Surprisingly, this department, which routinely concentrated on weapons production to the detriment of health and environmental concerns, still holds the purse strings for epidemiological research, and refuses to release all relevant health and disease data. DOE is so refractory that it recently announced that it will not collect any radiation worker data for analysis by

Health and Human Services, and it refuses to survey its own facilities to prepare an accurate inventory of its own data.[54] All large data bases of exposed populations should have ten independent stewards to ensure that data is not corrupted or interferred with by vested interests. These populations would include DOE workers, nuclear shipyard workers, commercial nuclear power plant workers, Hiroshima, Nagasaki, and Chernobyl victims, and American, Australian, Namibian, and East German uranium miners.

7. If necessary, whole cities may have to be evacuated, e.g., Denver, which is contaminated with plutonium, because of the extreme toxicity of the material. Of course there is no way that Denver and its suburbs can be decontaminated.

8. New radiation exposure standards from five rems per year to two rems per year must urgently be implemented for workers handling radioactive materials and waste (this work will be extremely dangerous). Even this dose is not safe, because revision of the atomic bomb data finds that radiation is at least three to four times more hazardous than previously understood. But the DOE refuses to implement this new level,[55] presumably because many more workers will now be required for the cleanup and operation of nuclear reactors because the dose legally allowed will be a small fraction than before.

9. All people with cancers or diseases related to U.S. government radiation exposure must be appropriately compensated, although how does one evaluate a life? This includes compensation of families who have lost relatives.

10. The radioactive waste must be defined according to its longevity, its toxicity, its biological pathways, and its predisposition to spread from its original source. Presently, there are three artificially defined categories of radioactive waste, which depend upon the mode of production: (a) high-level waste, which is spent reactor fuel and waste from reprocessing; (b)

transuranic waste contaminated with plutonium and its very toxic relatives, curium, neptunium, Americium, etc., all alpha emitters with very long half-lives; (c) low-level waste includes anything that doesn't fit the above categories. This can and often does contain plutonium. But the average radioactivity of commercial nuclear power plant "low-level" waste is three times higher than high-level waste from weapons production.[56] All three categories contain long-lived isotopes, like plutonium 239 with a radioactive life of 500,000 years; cesium 137, with a radioactive life of 600 years; strontium 90, also with a radioactive life of 600 years, etc. But the Nuclear Regulatory Commission has ruled that low-level waste need only be monitored for 100 years!

11. A new department of radioactive waste disposal must be created, staffed by the best scientists from DOE, Department of Defense, and from weapons contractors. Until recently, two-thirds of American scientists were creating nuclear weapons and delivery systems. A decision must be made of 100 percent containment and zero release for the entire hazardous life of the radioactive material, and the program must be planned to achieve this. These proposals will cost less than the $300 billion spent annually by the Pentagon and the DOE. This is true national security—ensuring the safety of U.S. citizens.

And remember, we did not inherit the earth from our ancestors, we borrowed it from our descendants.

Notes

INTRODUCTION

1. Australian Department of Trade Report. Canberra, 1985–1986.
2. Bill Birnbauer, "Uranium Sales to Indonesia Are Possible," *The Age*, August 30, 1993.
3. Keith Schneider, "Expert Warned Over Radiation Tests on Humans," *The Age*, December 29, 1993; Phillip McCarthy, "Poisoning Their Own, U.S. Experiments on Citizens to Keep Cold War Edge," *The Age*, January 1, 1994; "Clinton Holds Urgent Talks on Victims of Radiation Experiments," *Guardian*, Reuter, *The Age*, January 5, 1994; Keith Schneider, "Energy Official Seeks to Assist Victims of Tests," *New York Times*, December 29, 1993; "Paper Says Experiment Exposed 19 Retarded Youths to Radiation," *New York Times*, December 27, 1993; Matthew L. Wald, "Inquiry into Radiation Tests Extends to Veterans," *New York Times*, December 30, 1993; Anthony Lewis, "Secrecy and Cynicism," *New York Times*, December 27, 1993; Keith Schneider, "Secret Nuclear Research on People Comes to Light," *New York Times*, December 17, 1993; "America's Nuclear Secrets," *Newsweek*, December 27, 1993.

4. Melissa Healy, "US Reveals 204 Nuclear Tests, Plutonium Exposure," *Los Angeles Times,* December 8, 1993.

5. "Pilots Dropped Radiation in US," AP, *The Australian,* December 16, 1993.

6. Matthew L. Wald, "Rusting Uranium in Storage Pools Is Troubling US," *New York Times,* December 8, 1993.

7. Matthew L. Wald, "Explosions Feared at Arms Factories," *New York Times,* December 10, 1993.

8. "Russian Fleet Bid to Dump More Nuclear Waste," Reuter, *The Age,* December 22, 1993.

9. Christopher Reed, "Nuclear Plant Faces Safety Suit," *The Age,* January 10, 1994.

Chapter 1
OUR OWN WORST ENEMY

1. Christopher Burns, "Nuclear Cash Fallout," *Herald Sun,* June 14, 1993.

2. Nick Lenssen, "Vital Signs," World Watch Institute. (New York: W.W. Norton, 1993).

3. Helen Caldicott, *Missile Envy* (New York: Bantam, 1986), 295.

4. Greg Myre, "S. Africa Concedes Having Once Built Nuclear Weapons," *The Idaho Statesman,* March 25, 1993; Gary Milhollin, "The Iraqi Bomb," *The New Yorker* (Feb. 1, 1993); Seymour Hersh, "On The Nuclear Edge," *The New Yorker* (March 29, 1993).

Chapter 2
RADIATION

1. Hatsumi Nagasawa and John B. Little, "Induction of Sister Chromatid Exchanges by Extremely Low Doses of Alpha Particles," *Cancer Research* 52, (Nov. 15, 1992): 6394–6396. Dr. Howard Hu, Dr. Arjun Makhijani, Dr. Katherine Yih, *Plutonium: Deadly Gold of the Nuclear Age* (Cambridge Mass.: International Physicians Press, 1992).

2. Personal communication with Mary Olson, Nuclear Information Resource Services (NIRS), Washington, D.C.

3. BEIR Committee, Dec. 19, 1989, *Health Effects of Exposure to Low Levels of Ionizing Radiation* (Washington, D.C.: National Academy Press, 3101 Constitution Ave. NW); and John W. Gofman, "Radiation Induced Cancer From Low-Dose Exposure" (Committee for Nuclear Responsibility, P.O. Box 11207, San Francisco, CA, 1990).

Chapter 3

THE CYCLE OF DEATH

1. Keith Schneider, "A Valley of Death for Navajo Uranium Miners," *New York Times,* May 3, 1993.
2. Patricia Kahn, "A Grisly Archive of Key Cancer Data," *Science* 259 (Jan. 22, 1993).
3. Australian Department of Trade Report, 1985–86, Canberra; and Lindsay Murdoch, "Indonesia Puts Off Canberra N Deal," *Sydney Morning Herald,* Nov. 18, 1992.
4. "Lucas Heights Dirtier Than Overseas Reactors: Report," *The Age,* June 8, 1993.
5. Eric Hoskins, "Making the Desert Glow," *New York Times,* Jan. 21, 1993.
6. Briefing Paper on U.S. Nuclear Regulatory Commission 1990 Radiation Standards, NIRS, 1424 16th St. NW, Suite 601, Washington, D.C.
7. Personal communication with Greg Minor, MHB Technical Associates, San Jose, CA, Nov. 1992.
8. Matthew L. Wald, "U.S. Says 15 Reactors Need Testing," *New York Times,* April 1, 1993.
9. "Advanced Reactor Study," MHB Technical Associates, Consultants on Energy and the Environment, 1723 Hamilton Ave., Suite K, San Jose, CA, 3–8, 3–37.
10. Personal communication with Greg Minor, MHB Technical Associates, Nov. 1992.
11. Ibid.
12. Nuclear Regulatory Commission, *Information Digest,* 1991, 26.
13. Personal communication with Dale Bridenbaugh, MHB Technical Associates, Sept. 1992.

14. Matthew L. Wald, "NRC Reacts to 'Nuclear Targets' in Trade Center Letter," *New York Times*, March 26, 1993.

Chapter 4
NUCLEAR SEWAGE

1. Personal communication with Greg Minor.

2. Ibid.

3. Ibid.

4. "Russia's Nuclear Nightmare," AP, *The Northern Star*, July 22, 1992; Lester Brown et al., "State of the World,' A Worldwatch Institute Report on Progress Toward a Sustainable Society (New York: W.W. Norton, 1990); Alfred Friendly, "Warning: The Former USSR is Hazardous to Your Health," *Washington Post*, May 10, 1992.

5. "Russia Pays Dearly For its Ecological Sins," *Sydney Morning Herald*, Oct. 9, 1992.

6. Thomas B. Cochran and Robert Standish Norris, *Russia/Soviet Nuclear Warhead Production*, Natural Resources Defense Council, 1350 New York Ave. NW, Washington, D.C., May 15, 1992, 34.

7. *Long-Lived Legacy: Managing High-Level and TRU Waste at the DOE Weapons Complex*, Congress of the United States Office of Technological Assessment, OTA-BP-O-83, Washington, D.C., U.S. Government Printing Office, May 1991, 54, 66, 67.

8. Kay Drey, "Nuclear Power's Dirty Secret," *Viewpoint*, SECC, 1717 Mass. Ave, NW, Suite LL 215, Washington, D.C.

9. Personal communication with Dale Bridenbaugh, Sept. 1992.

Chapter 5
PLUTONIUM

1. Richard Tanksley, "Plutonium Radiation Could Cause 'Delayed Mutations,' " *New Mexico Daily Lobo* 92, no. 107, March 2, 1992.

2. Lobo Peace Group, "The Real Danger Lurks in the Airborn Dust," *New Mexico Daily Lobo* 96, no. 102, Feb. 24, 1992.

3. Ibid.

4. Petition of Objection: "Against the Transport of Plutonium From France," Oct. 1992, Plutonium Free Future, 2018 Shattuck Ave., Berkeley, CA 94204.

5. *Advanced Reactor Study,* 2–38.

6. Petition of Objection; Stansfield Turner and Thomas Davis, "Plutonium Terror on the High Seas," *New York Times,* April 28, 1990; Jacob M. Schlesinger, "Japanese Plan for Stockpiling Plutonium Draws Fire from Environmental Groups," *Wall Street Journal,* Oct. 29, 1991; "Sea Collision Reported in Japanese Chase," *Sydney Morning Herald,* Nov. 10, 1992; and "Japanese Plutonium Ship May Pass Australia," *Sydney Morning Herald,* Oct. 7, 1992; "The Plutonium Factor," Australian Broadcasting Corporation Radio National, July 19, 1992.

7. Personal communication with Dale Bridenbaugh, Sept. 1992.

8. "British Paper 'Foiled' Nuke Bid," *Sydney Morning Herald,* Nov. 2, 1992.

9. Personal communication with F. Berkhout, pending publication of D. Albright, F. Berkhout, and W. Walker, *A World Inventory of Plutonium and Highly Enriched Uranium, 1992* (Oxford University Press, forthcoming).

10. Robert Milliken, "Australia's Nuclear Graveyard," *Bulletin of the Atomic Scientists* (April 1987).

11. Maryann Stenberg, "Maralinga Radiation Survey Rises Tenfold," *The Age,* June 10, 1993.

12. Grigori Medvedev, *The Truth about Chernobyl* (New York: Basic Books, 1991), 224.

13. Letter from Dr. Paul Wolf, Northwestern University Library, Evanston, Ill., Oct. 19, 1986.

14. Steven Aftergood, David W. Hafemeister, Oleg F. Prilutsky, Joel R. Primack, and Stanislaw N. Rodionov, "Nuclear Power in Space," *Scientific American* (June 1991).

Chapter 6
M.A.D.: MUTUALLY ASSURED DESTRUCTION

1. *National Energy Strategy*, 1st ed., 1991/2, Washington, D.C., Feb. 1991, National Technical Information Service, U.S. Dept of Commerce, 5285 Port Roual Rd., Springfield, VA 22161, 24, 25.

2. Geoffrey Stevens, "Weapons the U.S. Pretended Not to See. A Review of the Sampson Option: Israel's Nuclear Policy and American Foreign Policy," *Toronto Star*, Nov. 16, 1991; Joel Brinkley, "Book on Israel Atom Arms Goes Beyond U.S. Estimates," *New York Times*, Oct. 20, 1991.

3. Reuter, "Iran Nuclear Bomb Fears," *Northern Star*, Nov. 19, 1992.

4. "Pakistan Admits to A Bomb Capability," *New York Times, Washington Post*, and *Sydney Morning Herald*, Feb. 10, 1992; and David Albright and Mark Hibbs, "North Korea's Plutonium Puzzle," *Bulletin of the Atomic Scientists*, Nov. 1992.

5. R. Dennis Hayes, "Eastern Europe's Nuclear Window," *The Nation* (Aug. 26/Sept. 2, 1991).

6. AP, "President Declares South Korea Nuclear Free," *Sydney Morning Herald*, Dec. 19, 1991; "North Korea's Plutonium Puzzle."

7. Charles Aldinger, "Korea Nuclear Free Halts U.S. Troop Cuts," *Sydney Morning Herald*, Oct. 19, 1992.

8. Reuter, "U.S. Explores North Korea Nuclear Sanctions," *The Age*, July 10, 1993.

9. Warren Osmond, "Disarmament Dream Becomes a Nightmare," and Ben Hills, "North Korea's Threat," *Sydney Morning Herald*, April 24, 1993; David E. Sanger, "Atomic Energy Agency Asks UN to Move Against North Koreans," *New York Times*, April 2, 1993.

10. David E. Sanger, "Wary of North, Seoul Debates Atomic Bomb," *New York Times*, March 19, 1993; Alan Freeman, "New Treaty Lets Canada Sell Uranium to Taiwan," *Globe and Mail*, March 8, 1993; Elaine Sciolino, "Christopher Signals a Tougher U.S. Line Toward Iran," *New York Times*, March 31, 1993.

11. Personal communication with Michael Mariotte of NIRS.

12. David Albright and Mark Hibbs, "Iraq and the Bomb: Were They Even Close?" *Bulletin of the Atomic Scientists* (March 1991).

13. Graham Barrett, "Major Widens Arms-to-Iraq Enquiry," *Sydney Morning Herald,* Nov. 18, 1992.

14. "Nuclear Bomb Designs Unsecured for a Week," *Arizona Daily Star,* Sept. 17, 1991.

15. Tony Hewitt, "Aust Heavy Military Sales Boom," *Sydney Morning Herald,* June 30, 1992.

16. "U.S. Top Supplier of Third World Arms," *New York Times, Sydney Morning Herald,* Aug. 13, 1991.

17. Lee Feinstein, "Arms R Us," *Bulletin of the Atomic Scientists* (Nov. 1992).

18. "U.S. Losing Track of Military Aid," *New York Times, The Age,* Dec. 28, 1991; "Reducing Third World Armies," *Sydney Morning Herald,* Sept. 10, 1991.

19. Walter H. Corson, ed., *The Global Ecology Handbook: What You Can Do About the Environmental Crisis* (Boston: Beacon Press, 1990) 42, 53.

20. Ruth Leger Sivard, "World Military and Social Expenditures, 1977." WMSE Publications, 1977.

21. United States Nuclear Regulatory Commission, *Code of Federal Regulations: Energy,* Rules and Regulations, title 10, chapter 1, 60.

22. Thomas B. Cochran and Robert Standish Norris, "Russia/Soviet Warhead Production," Natural Resources Defense Council, Washington, D.C., May 15, 1992.

23. Personal communication with Captain James Bush, Center for Defense Information, 1500 Mass. Ave NW, Washington, D.C., 20005.

24. Caldicott, *Missile Envy,* 30.

25. Personal communication with Captain James Bush, Center for Defense Information.

26. Alan Cooperman, "Ukraine not Playing Nuke Games," *Daily Camera,* April 7, 1993.

27. Personal communication with Captain James Bush, Sept. 1992.

28. Douglas Jehl, "U.S. Said to Drop Plan for Nine Tests of Nuclear Arms," *New York Times,* June 30, 1993; Mark Matthews, "U.S. Leans Towards Ban on Further Arms Tests," *The Age,* June 30, 1993.

29. Douglas Jehl, "Clinton Expected to Order Renewal of Nuclear Tests," *New York Times,* May 15, 1993.

30. Thomas W. Lippman, "For Utah Fallout Victims, Money is of Little Comfort," *Washington Post,* May 17, 1993; Carol Gallagher, *America Ground Zero* (Cambridge: MIT Press, 1993).

31. Caldicott, *Missile Envy,* 90.

32. Personal communication with Captain James Bush.

33. Patrick E. Tyler, "U.S. Strategy Plan Calls For Ensuring No Rivals Develop," *New York Times,* March 8, 1992.

34. "U.S. Urged to Target 5000 Missiles," *Sydney Morning Herald,* Jan. 7, 1992.

35. John Lancaster, "U.S. Should Aim for Dual War Capability," *Sydney Morning Herald,* June 26, 1993.

36. Personal communication with Ian Masters of KPFK, Los Angeles.

Chapter 7

THREE MILE ISLAND

1. Personal communication with Greg Minor.

2. Ibid.

Chapter 8

CHERNOBYL

1. André Carothers, "Children of Chernobyl," *Greenpeace* (Jan/Feb 1991), 10, 11.

2. Grigori Medvedev, *The Truth About Chernobyl* (New York: Basic Books, 1991), 32.

3. John W. Gofman. *Radiation-Induced Cancer From Low-Dose Exposure: An Independent Analysis,* 1st ed. (Committee for Nuclear Respon-

sibility, Inc, CNR Book Division, P.O. Box 11207, San Francisco, CA 94101), 16.

4. "Chernobyl Update," KPFK, Los Angeles, April 30, 1991.

5. *Radiation-Induced Cancer From Low-Dose Exposure,* 36–28.

6. Ibid., 5.

7. Ibid., 24–8.

8. Ibid., 24–5.

9. Ibid., 24–4.

10. Ibid., 25–16.

11. Ibid., 24–1.

12. National Energy Strategy, 1st ed., 1991–1992 (U.S. Government Printing Office, Washington, D.C., February 1991), 24–25.

13. Michael Parks, "Soviets Mark 26 Billion For Chernobyl Aid," *Los Angeles Times,* April 23, 1990.

14. "Inuit Food Contaminated," *World Perspectives,* vol. 3 (March/April 1992), 2, Box 3074, Madison, WI, 53704.

15. Douglas Stanglin et al., "Toxic Wasteland," *U.S. News and World Report* (April 13, 1992).

16. "Children of Chernobyl," 10, 11.

17. Cumbrians Opposed to a Radioactive Environment, Sept. 1986, 80 Church St, Barrow-in Furness, Cumbria, LA142HJ.

18. Yevgeniya Albats, "The Big Lie," *Moscow News,* Oct. 15, 1989.

19. Michael Parks, "Soviets Mark 26 Billion For Chernobyl Aid," LA Times, April 23, 1990.

20. "Chernobyl—Now a Move to Try Gorbachev," *Sydney Morning Herald,* Dec. 13, 1991, & Alfred Friendly Jr., Warning: The Former Soviet Union is Hazardous to Your Health, *Washington Post,* May 10, 1992

21. "Where Nuclear Disaster Waits to Happen," *Sydney Morning Herald,* July 6, 1992.

22. Matthew L. Wald, "Former Soviet Lands Still Lag in Nuclear Safety, U.S. Finds," *New York Times,* April 24, 1992.

23. "Russia Pays Dearly for its Ecological Sins," *Sydney Morning Herald,* Oct. 9, 1992.

24. Serge Scherman, "Atom Plant Leaks Radioactive Gases in a Russian Town," *New York Times,* March 25, 1992

25. "Chernobyl Inactive," *Sydney Morning Herald,* May 28, 1992.

26. Marlise Simons, "Russia Says It Needs to Keep Using Reactors," *Sydney Morning Herald,* June 23, 1993.

27. "Russia To Resume Its Nuclear Program," *Sydney Morning Herald* June 3, 1992.

28. "Children of Chernobyl," 8.

29. Walter H. Corson, ed, *The Global Ecology Handbook: What You Can Do About The Environmental Crisis* (Boston: Beacon Press, 1990).

30. "Children of Chernobyl," 11.

31. U.S. Keen to Find Jobs for Soviet Scientists, *Sydney Morning Herald,* Feb. 10, 1992.

32. "Iran in Deal for Nuclear Weapons," *Sydney Morning Herald,* Oct. 13, 1992.

33. "Radio National," Australian Broadcasting Corporation, Oct. 19, 1992.

34. Letter from Joshua Handler, Research Coordinator Of Nuclear Free Seas Campaign, to George Bush, 1436 U St. NW, Washington, D.C., 20009, April 7, 1992; Patrick E. Tyler, "Soviets' Secret Nuclear Dumping Causes Worry For Arctic Waters," *New York Times,* April 30, 1992; Hal Benton, "Russian Revelations Indicate Arctic Region is Awash in Contaminants," *Washington Post,* May 17, 1993; William J. Broad, "Soviets Lied Over Nuclear Dumps, Say Russians," *The Age,* April 28, 1993.

36. "Halt Nuclear Dumping, Japan Tells Russia," *The Age,* October 19, 1993.

35. Miro Certnetig, "A Shadow Over Siberia," *Globe and Mail,* March 6, 1993.

36. William J. Broad, "Russians Call Sub a Nuclear Danger," *New York Times,* May 2, 1993.

37. "The Soviet Union's Dirty Secrets," *Globe and Mail,* Toronto, November 10, 1992.

38. Alfred Friendly Jr., "Warning: The Former USSR is Hazardous to Your Health," *Washington Post,* May 10, 1992; "Russia's Nuclear Nightmare," *Northern Star,* July 22, 1992; John Hallam, "What You Weren't Told About Tomsk," Friends of the Earth, Sydney, Australia.

39. Thomas B. Cochran and Robert Standish Norris, *"Russia/Soviet Warhead Production,"* Natural Resources Defense Council, Washington, D.C., May 15, 1992.

Chapter 9
ADVANCED NUCLEAR REACTORS FOR THE UNITED STATES

1. *Advanced Reactor Study,* MHB Technical Associates, Consultants on Energy and the Environment, 1723 Hamilton Ave, Suite K, San Jose, CA 95125, 1.
2. *Natural History,* Jan. 1992.
3. *Readers Digest,* Sept. 1991.
4. Ibid.
5. "Reactors Add No New Energy For 10 Years, Not Man Apart," Friends of the Earth, Nov. 1974; Council on Scientific Affairs, "The Medical Perspective on Nuclear Power," *Journal of the American Medical Association,* Nov. 17, 1989.
6. National Energy Strategy, 1st ed., 1991/92, Washington, D.C., Feb. 1991, National Technical Information Service, U.S. Department of Commerce, 5285 Port Royal Rd., Springfield, VA 22161, and personal communication with Greg Minor, MHB Technical Associates, San Jose, CA.
7. Al Gore, *Earth in the Balance,* 327–28.
8. *Advanced Reactor Study.*
9. Ibid., 2–13
10. Ibid., 2–15.
11. Ibid., 2–22.
12. Ibid., 3–46.
13. Ibid., 2–23; ibid., 2–28, 2–29; ibid., 2–35.
14. Ibid., 2–36, 2–37.
15. Ibid., 7–13.
16. Ibid., 7–12.
17. Ibid., 3–40.
18. Ibid., 5–11.
19. Personal communication with Greg Minor of MHB Associates.

20. Walter Sullivan, "Plutonium Found in Plants' Roots," *New York Times*, Oct. 11, 1974.

21. "Japanese Anti-Nuclear Groups Warn Over 'Plutonium Ship' " *Northern Star*, Aug. 27, 1992.

22. *Plutonium Processing: Gearing up for Complex-21: A Primer on Plutonium Operations in the Nuclear Weapons Complex* (Energy Research Foundation, 537 Harden St., Columbia, SC 29205, March 1992), 13, 14.

23. William J. Broad, "Russia, U.S. to Test Nuclear Reactor," *San Francisco Chronicle*, April 6, 1993.

24. *Advanced Reactor Study*, 2–38.

25. Ibid., 2–48.

26. Ibid., 2–51.

27. Ibid., 3–9.

28. Ibid., 3–10.

29. Ibid., 3–11.

30. Ibid., 3–12.

31. 3–30, 3–31.

32. Ibid., 3–5.

33. Ibid., 3–37; ibid., 3–8.

34. Ibid., 4–12, 4–6.

35. Ibid., 1–8.

36. Ibid., 2–57.

37. Ibid., 2–57, 58, 59.

38. Ibid., 2–61, 62, 63.

39. Ibid., 2–63.

40. Ibid, 0–8; "House Passes Energy Bill," *The Nuclear Monitor*, 7, no. 19 (June 1, 1992), Nuclear Information and Resource Service, 1424 16th. St. NW, Suite 601, Washington, D.C. 20036.

41. Ibid., 5–6.

42. "Nuclear Power and National Energy Strategy," Nuclear Information and Resource Service, April 1991.

43. Matthew L. Wald, "Barriers are Seen to Reviving the Nuclear Industry," *New York Times*, Oct. 8, 1990; Matthew L. Wald, Cheap and Abundant Power May Shutter Some Reactors," *New York Times*, April 14, 1992.

44. Valerie Lee, "Nuclear Market in Asia Set to Surge," *The Sunday Age,* May 30, 1993.

45. E. Michael Blake, Jon Payne, Gregg Taylor, and Nancy J. Zacha, "Boston Attendees Examine the Value of the Atom," *Nuclear News* (Aug. 1992), 89, 90.

46. Kai Erikson, "Radiation's Lingering Dread," *Bulletin of the Atomic Scientists* (March 1991),

47. "Five-Day Summer Workshops Still Popular With Teachers," *American Nuclear Society Newsletter* (Sept. 1992).

48. "Nuclear News," *Nuclear Information and Resource Service* (Aug. 1992), 89, 90, 91.

49. Janice Brady, "Operating a Nuclear Visitors Center," *Nuclear News* (Aug. 1992), 45–47.

50. Matthew L. Wald, "Federal Inspector Criticizes Indian Point Nuclear Plant For Lapses on Safety," *New York Times,* April 21, 1993.

Chapter 10
THE COLD WAR ENDS AND THE HOT WAR STARTS

1. William Boesman and Christopher Dodge, *Nuclear Weapons Complex: Modernization and Safety,* Congressional Research Service Issue Brief (Washington, D.C.: The Library of Congress, Science Policy Research Division, Sept. 28, 1991).

2. *Dead Reckoning: A Critical Review of the Department of Energy's Epidemiological Research* (PSR Task Force on the Health Risks of Nuclear Weapons Production, 1000 16th St. NW, Suite 810, Washington, D.C. 1992), 17; "Assessing the Military's War on the Environment," *State of the World* (1991), 146, 147; John H. Cashman, Jr., "Report Lists 45,000 Potential Radioactive Sites," *New York Times,* April 9, 1992.

3. William Arkin and Robert S. Norris, *Taking Stock* (Greenpeace/NRDC, 1992) 1; and Robert Standish Norris, *Russian/Soviet Nuclear Warhead Production* (National Resources and Defense Council, 1350 New York Ave., NY, May 15, 1992).

4. *Complex Cleanup: The Environmental Legacy of Nuclear Weapons*

Production, Congress of the United States, Office of Technological Assessment, OTA-O–484, Washington, D.C., U.S. Government Printing Office, Feb. 1991, 3.

5. Ibid., 3, 4.

6. Ibid., 7.

7. Ibid., 7.

8. Ibid., 7.

9. Ibid., 7, 8, 9.

10. Helen Caldicott, *Missile Envy* (New York: Bantam, 1986).

11. Laurence Joliden, "New Name, New Mission for 'Star Wars,' " *USA Today,* May 14, 1993.

12. "Plutonium Processing: Gearing Up for Complex-21: A Primer on Plutonium Operations in the Nuclear Weapons Complex" (Energy Research Foundation, 537 Harden St., Columbia, SC 29205, March 1992) 8.

13. *Facing Reality: The Future of the U.S. Nuclear Weapons Complex* (Nuclear Safety Campaign, 1914 North 34th St., Suite 407, Seattle, Wash., May 1992), 7.

14. Thomas W. Dowler and Joseph S. Howard II, "Potential Uses for Low-Yield Nuclear Weapons in the New World Order: Roles and Missions", Speech, Los Alamos National Laboratory, 1992.

15. Laurence Joliden, "New Name, New Mission for 'Star Wars'," *USA Today,* May 14, 1993; Robert L. Park, "Star Warriors and Sky Patrol," *New York Times,* March 25, 1992.

16. Ibid., 7.

17. *Facing Reality,* 6, 7.

18. *Dead Reckoning,* 46.

19. Ibid., 39.

20. *Complex Cleanup,* 24.

21. *Complex Cleanup,* 157.

22. "Plutonium Processing", 17–24.

23. Ibid., 30.

24. *Russian/Soviet Nuclear Warhead Production,* 13.

25. "Plutonium Processing," 31; *Complex Cleanup,* 155.

26. Thom Cole and Kelly Richmond, "The $2 Billion Mess at LANL," *Santa Fe New Mexican,* Feb. 17, 1991.

27. John Stroud, "LANL Air Emissions Woes," *The Radioactive Rag* 4, no. 1 (Winter/Spring 1992), 412 W. San Francisco St., Santa Fe, NM.

28. "The $2 Billion Mess at LANL."

29. Keith Schneider, "Brain Cancer Cases in Los Alamos to be Studied for Radiation Link," *New York Times,* July 23, 1991.

30. *Dead Reckoning,* 46.

31. "Update on Environmental, Safety, Health, and Security Issues, Los Alamos National Laboratory," *New Mexican* insert, Oct. 3, 3.

32. *Missile Envy,* 258.

33. William Boesman and Christopher Dodge, "Nuclear Weapons Production Complex: Modernization and Safety," Congressional Research Service Issue Brief, The Library of Congress, Sept 23, 1991, 3.

34. *Complex Cleanup,* 166.

35. Kenneth B. Noble, "The U.S. for Decades Let Uranium Leak at Weapon Plant," *New York Times,* Oct. 15, 1988; *PSR Monitor,* vol. 6, no. 3 (Nov. 1990), 4; *Dead Reckoning,* 22.

36. Helen Caldicott, *If You Love This Planet* (New York: W.W. Norton, 1992), 90.

37. *Complex Cleanup,* 150.

38. "Nuclear Weapons Production Complex," 5; "Plutonium Processing", 43.

39. *Complex Cleanup,* 159, 160; *PSR Monitor,* vol. 6, no. 3 (Nov. 1990), 5.

40. *Dead Reckoning,* 46.

41. *Complex Cleanup,* 153, 154; *PSR Monitor,* vol. 6, no. 3, 4.

42. Ibid., *Complex Cleanup,* 162–163; *Dead Reckoning,* 19.

43. *Facing Reality,* 13.

44. *Complex Cleanup,* 156, 157.

45. *Dead Reckoning,* 24; *PSR Monitor,* vol. 6, no. 3, 5.

46. "Plutonium Processing", 42, 43.

47. *Complex Cleanup,* 152, 153.

48. *PSR Monitor,* vol. 6, no. 3, 4.

49. "Plutonium Processing," 25, 26.

50. *Dead Reckoning,* 22.

51. "Plutonium Processing," 27; Brian Abas, "Rocky Flats: A Big Mistake from Day One," *Bulletin of the Atomic Scientists* (Dec. 1989), 21.

52. Ibid., 26.

53. *Complex Cleanup*, 80.

54. Ibid., 18.

55. Ibid., 28.

56. "Rocky Flats, A Big Mistake from Day One."

57. *Complex Cleanup*, 165.

58. Tim Connor, "Nuclear Workers at Risk," *Bulletin of the Atomic Scientists* (Sept 1990), 26.

59. *Dead Reckoning*, 9.

60. Ibid., 46.

61. "Plutonium Processing," 29.

62. Ibid., 26.

63. Rusnock Hoover, "Flats Drops all Bomb Work," *Daily Camera*, April 6, 1993.

64. Karen Dorn Steele, "Hanford: America's Nuclear Graveyard," *Bulletin of the Atomic Scientists* (Oct. 1989), 18.

65. Larry Lande, "Missing Hanford Documents Probed by Energy Department," *Seattle Post-Intelligencer*, Sept. 20, 1991.

66. Scott Saleska and Arjun Makhijani, "Hanford Cleanup: Explosive Solution," *Bulletin of the Atomic Scientists* (Oct. 1990), 19.

67. Barbara Goss Levi, "Hanford Leaks Short- and Long-Term Solutions to its Legacy of Waste," *Physics Today* (American Institute of Physics, March 1992).

68. "Hanford: America's Nuclear Graveyard."

69. *Complex Cleanup*, 151.

70. "Hanford: America's Nuclear Graveyard."

71. Personal Communication with MHB Technical Associates, 1723 Hamilton Ave., Suite K, San Jose, CA.

72. *Dead Reckoning*, 24.

73. "Plutonium Processing," 35.

74. Gerald Pollet, "Seeking Human Justice," *Facing Reality*, 31.

75. *Dead Reckoning*, 46.

76. Plutonium Processing, 35.

77. "Plutonium Processing," 36–41.

78. Brian Costner, "Tritium Operations," *Facing Reality*, 13.

79. Peter Gray, "Overview," *Facing Reality*, 1.

80. Gordon Thompson and Steven C. Sholly, "Let's X-out the K," *Bulletin of the Atomic Scientists* (March 1992), 14.

81. William Lanoutte, "Weapons Plant at 40: Savannah River's Halo Fades," *Bulletin of the Atomic Scientists* (Dec. 1990), 27.

82. "Long-Lived Legacy: Managing High-Level and Transuranic Waste at the DOE Nuclear Weapons Complex," Congress of the U.S. Office of Technological Assessment, Washington, D.C. Government Printing Office, May 1991, 26; *PSR Monitor*, vol. 6, no. 3, 4.

83. Ibid., "Long-Lived Legacy: Managing High-Level and Transuranic Waste at the DOE Nuclear Weapons Complex," 20.

84. "Assessing the Military's War on the Environment" 147.

85. *Complex Cleanup*, 166.

86. "Weapons Plant at 40."

87. *Dead Reckoning*, 46, 48.

88. "Plutonium Processing," 42.

89. *Russian/Soviet Nuclear Warhead Production*, 16; Robert Norris and William Arkin, "Nuclear Notebook: Pantex Lays Nukes to Rest," *Bulletin of the Atomic Scientists* (Oct 1992), 48, 49.

90. "Plutonium Processing," 12.

91. *Russian/Soviet Nuclear Warhead Production*, 17, 18, 19.

92. "Weapons Production Laborious," *Northern Star*, July 29, 1992.

93. "Russia Offers Uranium to West," *Sydney Morning Herald*, July 23, 1992.

94. "CIA Warns of Soviet Nuclear Mercenary Threat," *Sydney Morning Herald*, June 2, 1992.

95. "Plutonium Processing," 13.

96. Personal communication with Captain James Bush, Center for Defense Information, Washington, D.C.

97. *Complex Cleanup*, 160, 161.

98. *Dead Reckoning*, 48.

99. Ibid., 56.

100. "Long-Lived Legacy," 10.

101. "CIA Warns of Soviet Military Threat," *Sydney Morning Herald,* June 2, 1992.

102. "Tale of Hand-Built Soviet Arsenal," *Sydney Morning Herald,* Feb. 8, 1992.

Chapter 11

WASTE CLEANUP

1. "Whatever will we do with radioactive wastes?," *Peace News,* vol. 9, no. 1 (Winter 1993) The Colorado Coalition, 1738 Wynkoop, Suite 302, Denver, CO.

2. *Complex Cleanup,* 56.

3. Ibid., 39, 57.

4. "Radioactive Waste: Buried Forever," Radioactive Waste Campaign, 7, West St., Warwick, NY 10990.

5. Jonathan Becker, *Deregulating Radioactive Waste Disposal* (Public Citizen, Critical Mass Energy Project, 215 Pennsylvania Ave., SE, Washington, D.C., May 29, 1990).

6. Diane D'Arrigo, "NIMBY: Nukewaste in My Backyard," *Southwest Research and Information Center,* vol. 14, no. 2 (April/June 1989).

7. "Background Briefing on the Current Status of BRC," *Nuclear Information and Resource Service,* Aug. 20, 1992.

8. NIMBY.

9. John W. Gofman, *Radiation-Induced Cancer From Low-Dose Exposure:* An Independent Analysis, 1st ed., 1990 (Committee for Nuclear Responsibility, Inc., CNR Book Division, P.O. Box 11207, San Francisco, CA 94101), 1–3; *Health Effects of Exposure to Low-Levels of Ionizing Radiation,* BEIR FIVE, Committee on the Biological Effects of Ionizing Radiations, National Academy of Sciences, Washington, D.C., 1989.

10. NIMBY.

11. John W. Gofman, *Radiation and Human Health* (Sierra Club Books, 1981), 578–583.

12. Personal communication with Mary Olson, Nuclear Information Resource Services.

13. Ibid.

14. "Long-Lived Legacy, Managing High-Level and Transuranic Waste at the DOE Weapons Complex." Background Paper, Congress of the U.S. Office of Technological Assessment, U.S. Government Printing Office, May 1991, 24, 25.

15. *Deregulating Radioactive Waste Disposal.*

16. NIMBY.

17. "Long-Lived Legacy: Managing High-Level and Transuranic Waste at the DOE Weapons Complex, Background Paper, Congress of the U.S. Office of Technological Assessment, U.S. Government Printing Office, May 1991, 61, 67.

18. "DOE May Have Shipped Illegal BRC Waste to 26 Sites Nationwide," *The Nuclear Monitor,* Sept. 23, 1991.

19. Personal communication with Mary Olson, NIRS.

20. Ibid.

21. Ibid.

22. "Oklahoma Tribes Eye Waste Site," *Tulsa Tribune,* Jan. 9, 1992; David Leroy, speech to National Congress of American Indians, Office of the U.S. Nuclear Waste Negotiator, Boise, Idaho, Dec. 4, 1991.

23. Personal communication with Mary Olson, NIRS.

24. "Long-Lived Legacy," 59.

25. Ibid., 57.

26. Ibid., 56, 71.

27. Ibid., 72, 73, 58.

28. Matthew L. Wald, "The Adventures of the Toxic Avengers Have Only Just Begun," *New York Times,* Sept. 15, 1991.

29. "Waste Isolation Pilot Plant (WIPP) Fact Sheet," Concerned Citizens for Nuclear Safety, 412, West San Francisco St., Santa Fe, N.M., Feb. 19, 1991.

30. "Yucca Mountain and Nuclear Waste Transportation," Citizen Alert, P.O. Box 1681, Las Vegas, Nevada.

31. "Background Status of High-Level Nuclear Waste Management," Nuclear Information and Resource Service, Aug. 1992; "Stop the Yucca Mountain Nuclear Waste Dump," Greenpeace c/o Nuclear Information and Resource Service.

32. Ibid.

33. "Industry Mounts Massive Campaign to Cause Collapse of Antinuclear Forces in Nevada," *Nuclear Monitor,* vol. 7, no. 7, Dec. 1991.

34. James Coates, "A Jolt to a Nuclear Plan," *Boston Globe,* July 29, 1992.

35. "Yucca Mountain and Nuclear Waste Transportation."

36. "EPA Proposed High-Level Waste Standards," 40 Code of Federal Regulations, part 91, conference committee report; The National Energy Policy Act of 1992 Draft High Level Waste/Carbon 14 Release Subcommittee Report for consideration by the EPA Executive Committee (For estimate of C 14 to be released).

37. "Stop the Yucca Mountain Nuclear Waste Dump," Greenpeace.

38. "Yucca Mountain and Nuclear Waste Transportation."

39. Ibid.

40. Background and testimony of Gregory Minor on behalf of the Minnesota Department of Public Service, Docket No-002/CN-91-19, Sept. 1991.

41. "Long-Lived Legacy," 28.

42. Ibid., 27.

43. Ibid., 44.

44. Ibid., 42.

45. Ibid., 4.

46. Ibid., 30, 45.

47. Helen Caldicott, Missile Envy, 66–70; Keith Schneider, "Estimates of Weapons Cleanup Inflated," *New York Times,* April 30, 1992.

48. Matthew J. Wald, "The Adventures of the Toxic Avengers Have Only Just Begun," *New York Times,* Sept. 15, 1991.

49. "California: About to Become Site for Nuclear Dump?" Don't Waste California, 2940 16th St., #310, San Francisco, CA 94103.

50. Caroline Courtwright, "A Dump in the Desert: The Proposed Ward Valley Project in California," *Nuclear Guardianship,* Forum Issue (Spring 1992), 1400 Shattuck Ave., #41, Berkeley, CA; Larry Stammer and Paul Feldman, "Nuclear Dump Denounced at State Hearings," *Los Angeles Times,* July 23, 1991.

51. "1000 Frozen Dogs in Davis," *Sacramento Bee,* Sept. 4, 1990.

52. "Radioactive Waste: The Medical Factor," a report by Minard Hamilton, from the Nuclear Information and Resource Service, 1993.

53. Complex Cleanup: The Environmental Legacy of Nuclear Weapons Production, Congress of the United States, Office of Technological Assessment, OTA–484, Washington, D.C., U.S. Government Printing Office, Feb. 1991, 19.

54. "Facing Reality, The Future of the U.S. Nuclear Weapons Complex"; H. Jack Geiger and Daryl C. Kimball, "Health Effects of Nuclear Weapons Production": DOE Research, 19.

55. *Dead Reckoning, Physicians for Social Responsibility,* 1992, 1000 16th St. NW, Suite 810, Washington, DC 20036, 32.

56. "Facing Reality, The Future of the U.S. Nuclear Weapons Complex"; Don Hancock and Arjun Makhijani, Radioactive Waste Storage and Disposal, 22.

Index

"Helen Caldicott has been my inspiration to speak out."
—Meryl Streep

As a physician, I contend that nuclear technology threatens life on our planet with extinction. If present trends continue, the air we breathe, the food we eat, and the water we drink will soon be contaminated with enough radioactive pollutants to pose a potential health hazard far greater than any plague humanity has ever experienced.

First published in 1978, Helen Caldicott's *cri du coeur* about the dangers of nuclear power became an instant classic. In the intervening sixteen years much has changed—the Cold War is over, nuclear arms production has decreased, and there has been a marked growth in environmental awareness. But the nuclear genie has not been forced back into the bottle. The disaster at Chernobyl and the "incidents" at other plants around the world have disproven the image of "safe" nuclear power. Nuclear waste dumping has further poisoned our environment, and developing nuclear technology in the Third World poses still further risks.

In this completely revised, updated, and expanded edition, Dr. Caldicott defines for the 1990s the dangers of this madness—including the insidious influence of the nuclear power industry and the American government's complicity in medical "experiments" using nuclear material—and calls on us to accept the moral challenge to fight against it, both for our own sake and for that of future generations.

An internationally acclaimed Australian physician and antinuclear activist, Helen Caldicott was co-founder of Physicians for Social Responsibility, and founder of the Women's Action for Nuclear Disarmament and International Physicians to Save the Environment. She is also the author of *If You Love This Planet* and *Missile Envy*.

Cover design by Linda Kosarin

W·W·NORTON

NEW YORK · LONDON

ISBN 0-393-31011-6

90000 >

EAN

9 780393 310115

$10.95 USA $13.99 CAN.